T0329585

Laser Inter-Satellite Links Technology

IEEE Press
445 Hoes Lane
Piscataway, NJ 08854

Laser Inter-Satellite Links Technology

Jianjun Zhang
China Academy of Space Technology, Beijing, China

Jing Li
Beijing Institute of Technology, Beijing, China

Published by John Wiley & Sons, Inc., Hoboken, New Jersey.
Published simultaneously in Canada.

Library of Congress Cataloging-in-Publication Data applied for:
Hardback ISBN: 9781119910718

Cover Design: Wiley
Cover Image: © greenbutterfly/Shutterstock

Set in 9.5/12.5pt STIXTwoText by Straive, Pondicherry, India

Contents

Author Biography

Jianjun Zhang, PhD, Professor. He received his PhD degree from the Institute of Optoelectronics, Chinese Academy of Sciences, in 2010. He is now Professor at Beijing Institute of Spacecraft System Engineering, China Academy of Space Technology. He is also the member of the Youth Science Club of China Electronics Society, member of the Edge Computing Expert of China Electronics Society, Chairman of the "Space (Aerospace) Information Technology" Professional Committee of China Electronics Society, and member of the Satellite Application Expert Group of China Aerospace Society. He is mainly engaged in satellite navigation system design and advanced spatial information system technology based on cognitive mechanism. He has presided over several major projects such as the National Natural Science Foundation's major research project, the final assembly fund, the 863 project, and the development project of the Science and Technology Commission of the China Academy of Space Technology. He has published more than 50 SCI/EI search papers in international journals and conferences, authorized more than 20 invention patents at home and abroad, and published 3 monographs. He won the third prize of National Defense Science and Technology Progress Award with the first completion person.

Jing Li, PhD, Associate Professor, Supervisor. She received her PhD degree from the Beijing Institute of Technology in 2011. She is now Associate Professor at the School of Automation, Beijing Institute of Technology. She is an expert member of the "Space Information Technology" Youth Committee of China Electronics Society. Her main research direction is robot environmental awareness, image detection and target tracking, and multi-sensor information fusion. She has presided over more than 10 projects including the National Natural Science Youth Fund, the Postdoctoral Special Fund, the Key Laboratory of the Ministry of Education, and the Science and Technology Cooperation. She has published 25 academic papers (including 10 SCI papers and 15 EI papers) and the book "Image Detection and Target Tracking Technology", granted 7 national invention patents as the first author and the National Science and Technology Progress Award 2 (ranked 8th). The guided postgraduates have awarded the second prize of the 14th China Graduate Electronic Design Competition, the second prize of the first China-Russia (Industrial) Innovation Competition, and the second prize of the 14th National College Student Smart Car Competition.

Preface

In recent years, with the rapid growth of the demand for satellite applications in people's daily life, various needs are more and more inseparable from the navigation system, and the role of the satellite navigation system in people's daily life is becoming more and more important. The development of satellite navigation systems plays an important role in national economic construction and military defense work. Therefore, the world's major powers regard the construction of modern global navigation satellite system GNSS as a major strategic basic resource allocation. The four existing GNSS systems include GPS in the United States, Galileo in Europe, GLONASS in Russia, and Beidou in China, as well as other satellite navigation systems under construction or to be built in the future, and their related augmentation systems.

The normal operation of the ground station is an important support for the stable operation of the existing GNSS system. If the ground station is destroyed during combat or in other situations, the entire GNSS system will be paralyzed, and the consequences will be unimaginable, which will bring benefits to the entire country. The damage is irreparable. The inter-satellite link is also an important part of the GNSS system, which links the navigation satellites in the same orbit or between different orbits through transmitters, antennas, and receivers. Reducing the cost of system construction is one of the benefits brought by ISL technology. The inter-satellite link ranging and communication functions can effectively reduce the number of ground stations required by the system. Even if there is no ground station support, the system can still be used in cycles. Normal work is especially important for countries with limited resources and no global deployment capability. The Block-IIR series satellites in the second-generation GPS system of the United States are the only inter-satellite link systems with ranging and communication functions among all the GNSS navigation satellites currently in orbit. In addition, the construction of inter-satellite links, whether it is the third-generation GPS system or GNSS in other countries, is still in the stage of research and demonstration.

The International Telecommunication Union (ITU) has not yet allocated a laser band inter-satellite link, and the radio frequency link covers several radio frequency bands from ultra-high frequency (UHF) to extremely high frequency (EHF) allocated by the ITU. Compared with traditional inter-satellite radio frequency technology, satellite laser communication technology is known as a new generation of space communication technology. Laser carrier has some unique advantages: large communication capacity, good concealment performance, high bandwidth and transmission rate, strong anti-interference ability, no radio frequency license, etc., are required. At the same time, according to the modulation method and the detection method of the receiving end, the inter-satellite laser communication can be divided into incoherent – Intensity Modulation/Direct Detection (IM/DD) and coherent systems. The former uses light intensity modulation and direct detection schemes, which use precise modulation techniques (such as BPSK, QPSK, M-QAM, etc.) combined with coherent reception techniques (homodyne and heterodyne) schemes. Studies have shown that the IM/DD communication system cannot approach the theoretical limit of detection sensitivity in the actual inter-satellite link environment, and the sensitivity and wavelength selectivity of the coherent method are greatly improved compared to the IM/DD system.

China is promoting the construction of the Beidou second-generation satellite navigation system in an orderly manner, but for various reasons, China currently does not have global geographic strategic resources, and can only set up ground stations in China or parts of Asia. Therefore, the research on the inter-satellite link of the navigation system and the construction task is urgent and affects the overall work. Precise ranging and data transmission between navigation satellites is the premise for the system to achieve precise positioning and time synchronization. It can be seen that the application of laser technology to the inter-satellite link of the navigation constellation can enhance the anti-interference ability and confidentiality of the system at the same time, and what is more expected is to significantly improve the inter-satellite ranging accuracy and communication rate, thereby enhancing the survivability of the entire GNSS system. It is foreseeable that the application prospect of the inter-satellite link laser ranging and communication integration technology is immeasurable, so the research on the laser inter-satellite link technology is indispensable.

1 Introduction

1.1 Connotation of Inter-Satellite Link

With the rapid development of aerospace technology, national interests are gradually expanding beyond the traditional territory, territorial waters, and airspace, expanding and extending to the ocean, space, and electromagnetic space. With the rapid development of space technology, space has increasingly become a new source of international strategic competition commanding heights. Space technology embodies the political strength, economic strength, and scientific and technological strength of a country, and has become a strategic means for countries to demonstrate the progress of their national strength and defend their international status. In order to safeguard our maritime rights and interests, defend our space rights and interests, and ensure the expansion of national interests, China must focus on a global perspective and develop satellite navigation, communication, and other systems on a global scale. To this end, it is necessary to use space-based networking methods to break through the limitations of land and solve problems such as satellite full-orbit operation management, constellation autonomous operation, and rapid response to complex tasks, in order to ensure the development of space systems [1].

Satellite technology is the primary breakthrough for occupying space resources. After going through two stages of single-satellite application and constellation application, the satellite field has gradually moved toward networking. The networking between satellites must first require that the information and data between satellites can be interconnected. Considering factors such as security and national strategy, the number of satellite ground stations that can be established is very limited, and most of them are limited to the country. When the ground station is strictly limited, the establishment of an inter-satellite link has become one of the most important necessary conditions for inter-satellite networking [2, 3].

Laser Inter-Satellite Links Technology, First Edition. Jianjun Zhang and Jing Li.

Published 2023 by John Wiley & Sons, Inc.

Once a communication network is established between satellites through inter-satellite links, the satellites are no longer isolated individuals, but a whole with a considerable scale, and satellites can only complete a small number of tasks to complete a general a lot of work. Under a good inter-satellite link network, the satellite not only has more powerful functions, but also its robustness and anti-interference have also been greatly enhanced.

The constellation network is a very complex space network. This is because in the inter-satellite link network, there are not only a large number of space nodes, but also many ground nodes are also part of the network, the composition is complex and changeable, the access is flexible and irregular, with typical flat and centerless features. These features require a high degree of flexibility and adaptability to build inter-satellite networks. In addition, the inter-satellite link is not a single, pure communication link. While carrying a certain rate of communication tasks, it may also need to complete high-precision inter-satellite measurement functions at the same time. In the navigation system, the inter-satellite link needs to meet the core requirements of precise orbit determination and time synchronization of the system, autonomous navigation applications, etc., and support the use of satellite–ground joint orbit determination and the transmission capability of measurement data for autonomous navigation. The construction put forward higher requirements.

Inter-satellite link (or crosslink) refers to the link between satellites and can also be extended to the link between spacecraft. The inter-satellite link can perform functions such as inter-satellite communication, data transmission, inter-satellite ranging, and inter-satellite measurement and control. Different space systems have different functions of inter-satellite links. The inter-satellite link of the communication satellite constellation can reduce the number of satellite–ground hops and communication delay; the inter-satellite link of the reconnaissance formation system can increase the aperture of the virtual camera and improve the resolution; the inter-satellite link of the navigation satellite constellation can support autonomous operation and improve the resolution. Positioning accuracy, the inter-satellite link of the relay satellite system can increase the measurement and control arc of the user satellite. In addition, there are some inter-satellite links used in scientific research, such as gravity detection satellite systems. The inter-satellite link makes multiple satellites form an organic whole to form a constellation system and expand the ability of a single satellite to work [4].

At present, distributed satellite systems mainly include two types: formation satellite constellation and formation satellite. A typical constellation includes the sharing of scientific data through inter-satellite links between satellites in orbits of planets or the sun. They do not rely on each other to complete autonomous onboard navigation corrections, which are transmitted through ground stations.

The formation satellites need to rely on inter-satellite links to transmit navigation data to achieve fully autonomous navigation. At the same time, formation satellites also need to ensure strict positioning accuracy to meet the scientific goals of formation missions. One or several spacecraft in the formation have the function of navigation and processing, and maintain the formation or ensure a certain topology structure through the transmission of data and instructions. At the same time, the state information of the spacecraft also needs to be transmitted by inter-satellite links.

Figure 1.1 divides formation satellite systems into different classes from an inter-satellite link perspective. There are two kinds of communication links between formation satellites: satellite–ground link and inter-satellite link. If satellites transmit information through satellite–ground links, direct communication between spacecraft is generally not required, but data are collected and processed on the ground, and then integrated into scientific or navigational information to be transmitted back to the spacecraft. Usually, this method is very dependent on the ground and can be regarded as centralized control, that is, a star-shaped distributed system. Some constellations exchange and process data entirely through satellite–ground communication. Unlike constellations, formation satellites use inter-satellite communications to exchange navigation data and commands. In addition, the topological structure of the communication network of formation satellites is mainly divided into two types: star type and point-to-point type (Figure 1.1).

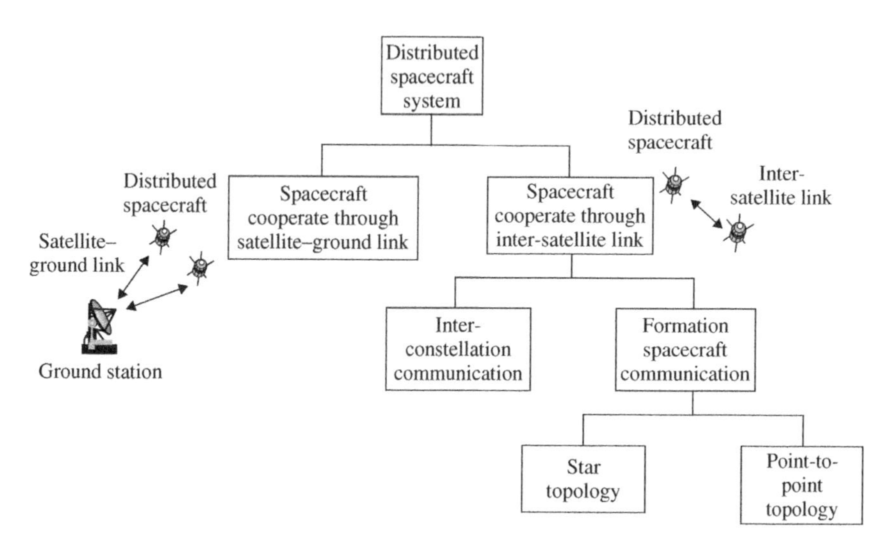

Figure 1.1 Distributed spacecraft communication structure division.

The use of inter-satellite links has many advantages, mainly including:

1) For communication satellites, when users who are not within the coverage area of the same satellite need to communicate, the use of inter-satellite links can eliminate satellite double hops and reduce the propagation delay. At the same time, a relay earth station dedicated to relaying signals between users in different satellite coverage areas is omitted [2, 5].
2) When the system is a constellation composed of many satellites, the use of inter-satellite links to form a complete communication network for all satellites is not only independent of the ground, but also greatly improves the system's anti-interference and anti-destroy capabilities.
3) It can be used to expand the coverage of the system. Multiple satellites are linked together through inter-satellite links, and users within the coverage area of any satellite can communicate directly with users within the coverage area of other satellites through the inter-satellite link.
4) Facilitate network management and form a global seamless network. For some low-orbit satellites, there may be no fixed earth station visibility at all in some cases (e.g. in the middle of the Pacific Ocean) due to their small coverage. At this time, it is almost the only solution to use the inter-satellite link to realize the control of the satellite on the ground and the mobile user to access the ground communication network through the inter-satellite link.
5) It is convenient to form a satellite group at the same orbital position, which is very useful for high-orbit satellites. Due to the increasing traffic volume of communications and the limited orbital positions of high-orbit satellites, it is bound to hope to make more effective use of each orbital position, and to place multiple satellites spaced about 100 km away from each other in one orbital position, so that they are interconnected by inter-satellite links to form a satellite constellation. This not only avoids the problem that a single satellite is too large to be loaded into orbit with existing launch tools, but also greatly reduces the risk of launch failure and satellite failure, and the capacity of the satellite can be adjusted according to actual needs. Incrementally, by increasing the number of satellites.
6) The space segment part of a system may be a constellation consisting of many satellites. For political and economic reasons, it is impossible to build a large terrestrial network that would allow any one of the satellites in the system to be able to see one of the ground control earth stations at any time. At this time, an inter-satellite link can be used to interconnect a satellite in the system that cannot be seen by the ground control earth station to another satellite (such as a high-orbit satellite) that can simultaneously see the ground control earth station and the satellite [6, 9].

On the other hand, the use of inter-satellite links also increases some design difficulties, including the need to increase the transceiver equipment necessary to maintain one or more inter-satellite links, including transceiver antennas, transceiver, radio frequency equipment, modulation and demodulation equipment, and the necessary baseband processing equipment. The satellite is required to have onboard processing function to distinguish whether the signal is sent to the inter-satellite link or the downlink. At the same time, it needs to have the necessary switching equipment to realize the onboard routing and exchange of the signal. Furthermore, the communication over the inter-satellite link should be transparent and the signal quality should not be degraded. All of these will inevitably increase the complexity of the satellite, and may increase the power burden of the satellite [7].

1.2 Types of Inter-Satellite Links

The type of inter-satellite link depends on the user with different usage requirements. However, its basic division includes two kinds: one is division by space domain and the other is division by frequency domain. Inter-satellite links are divided by frequency domain: ITU has allocated 14 frequency bands for inter-satellite links, ranging from UHF to EHF (190 GHz), including unallocated laser bands. That is, it can be summarized as microwave, millimeter wave, and laser links in three frequency bands. The inter-satellite link can be divided into the following two situations according to the airspace, that is, according to the satellite orbit [8–10]:

1) Inter-satellite links between satellites of the same orbit type, such as: GEO/GEO, LEO/LEO, etc.

   The inter-satellite links between satellites of the same orbit type are further divided into interstellar links in the same orbital plane and interstellar links in different orbital planes. Since the relative positions of the interstellar link satellites in the same orbital plane are fixed, we generally only analyze the different interplanetary link within the orbital plane. The interstellar links of the same orbit type mentioned in the subsequent chapters refer to the interstellar links in different orbital planes.

   Under the same coverage as shown in Figure 1.2, the GEO/GEO or MEO/MEO inter-satellite link can greatly improve the communication capacity of the system; when covering different areas, it can greatly increase the area of communication coverage; at the same time, it can improve the ground station. The minimum elevation angle can improve the quality of communication; it

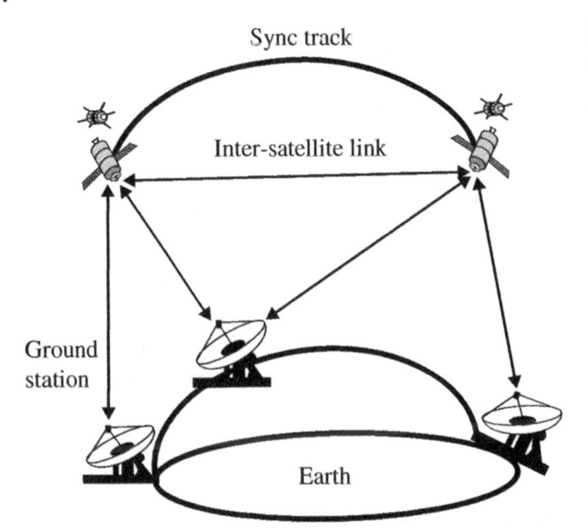

Figure 1.2 Schematic diagram of GEO/GEO inter-satellite link.

can also reduce the restrictions on satellite orbital positions and establish a global satellite communication network.

The LEO satellite inter-satellite link can make up for the two shortcomings of LEO satellite communication: one is that the coverage of a single satellite is very small, and the other is that the continuous communication time of a single satellite is very short, such as for Motoroh's iridium system, the maximum sustainable communication time of a single satellite is only 16 minutes. If the ground gateway station is used as a bridge for information communication between satellites, dozens or hundreds of ground stations may be required. Therefore, for the LEO system, it is unrealistic to rely on the ground station alone to communicate information between the satellites of the system. The inter-satellite link can realize the aerial networking of the satellite mobile communication system. In some cases, the satellite communication signal can reach the ground station only once or not at all before reaching the final communication user, thereby greatly saving the satellite mobile communication system. Invest in the ground segment and enable rapid transfer of information.

2) Interplanetary links between satellites of different orbit types, such as GEO/LEO, MEO/LEO inter-satellite links, also known as inter-orbital links (IOL).

Because the LEO satellite communication system requires a large-scale constellation (tens to hundreds) to form an effective global or regional coverage, the system investment is large, using one or more GEO or MEO satellites and LEO satellites to form a double layer. The satellite network can effectively overcome this disadvantage and maintain the advantages of the LEO system. In addition, the use of the GEO–LEO inter-satellite link can also form a tracking and communication support network for the spacecraft, with the ability to

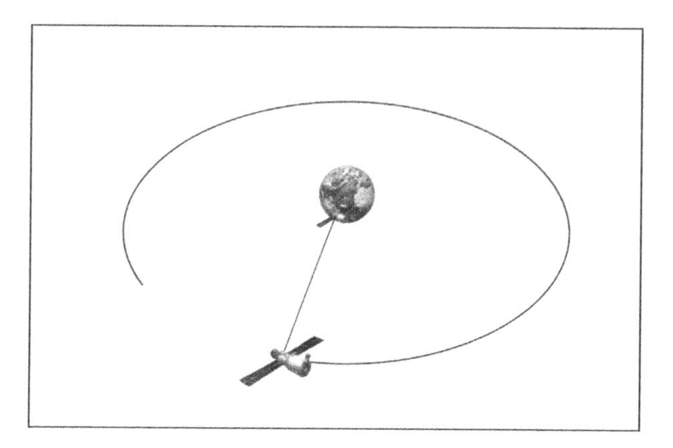

Figure 1.3 Schematic diagram of GEO–LEO interplanetary link.

provide services for multiple spacecraft at the same time; it can provide all-weather, full-orbit continuous tracking, orbit measurement and attitude measurement as shown in Figure 1.3.

1.3 Band Selection of Inter-Satellite Link

Inter-satellite links refer to satellite-to-satellite cross-links that can be used to replace the work done by terrestrial relay stations. The satellite link directly transmits information and data between satellites, avoiding the tedious work and frequent selection of routes such as sending the information back to the ground for processing and then back to the satellite, and also reduces the communication delay compared with the satellite–ground link. The purpose of the inter-satellite link is not only to achieve inter-satellite ranging, but also the high-speed, stable, and low-bit error rate transmission of inter-satellite data is one of its main purposes. For inter-satellite microwave communication, different data transmission rates need to select different inter-satellite link frequency bands, and high-speed data transmission must use high-frequency bands as carriers. Therefore, the selection of the inter-satellite link frequency band requires a compromise between hardware complexity and high-speed data transmission [11, 12].

1.3.1 Selection of Link Band

The inter-satellite link is a wireless link, and the communication and ranging of the inter-satellite link can use a microwave link or an optical link. When using a laser link (mainly a laser link), it has the characteristics of high measurement accuracy (can reach the sub-millimeter level), large communication capacity, good concealment and strong anti-interference ability, etc. However, the laser link device has a complex composition, the cost is high, the volume is large, the power consumption is high, the ranging distance is short, and scanning and alignment are required, which is not conducive to multi-link communication. These shortcomings limit the application of laser links to high ranging accuracy. When a microwave link is used, the bandwidth and antenna beam of the entire inter-satellite link are generally narrow, the frequency of the carrier itself is relatively low, and a large-sized antenna needs to be equipped on the satellite. The transmit power of the antenna is also high, the ranging accuracy is about from centimeters to decimeters, the link acquisition time is short, and multi-link communication is supported, which can be used for satellite constellations with a small amount of information transmission. At present, VHF and UHF frequency bands are mainly used. In recent years, SHF and EHF frequency bands have shown good application prospects due to their large transmission capacity, good spatial propagation characteristics, and no power density limitation. Moreover, since pseudo-code ranging can realize multiple measurements and spread spectrum communication, the pseudo-code ranging system is currently used in major long-distance space ranging at home and abroad, so microwave high-frequency bands have been paid more and more attention. In the GPSIII plan of the United States, Ka and V frequency bands will be used to meet the requirements of precise inter-satellite ranging and high-speed information transmission [13–15].

The commonly used microwave frequency bands are listed in Table 1.1. The frequency and band ranges in the table are both open on the left and closed on the right, that is, they contain an upper limit and do not include a lower limit.

1.3.2 Selection of Working Frequency

The frequency of the inter-satellite link of the satellite navigation constellation system is mainly divided into two major components: radio frequency and optical

Table 1.1 Composition of microwave frequency bands.

Band name	Frequency range	Band name	Wavelength range
UHF	300–3000 MHz	Decimeter wave	100–10 cm
SHF	3–30 GHz	Centimeter wave	10–1 cm
SHF	30–300 GHz	mmWave	10–1 mm
VHF	300–3000 GHz	smeepo	1–0.1 mm

Table 1.2 Frequency allocation of inter-satellite links.

Radio communications regulations	0.3–3 GHz 22.55–23.55 GHz 32–33 GHz 54.25–58.2 GHz 59–64 GHz 116–134 GHz 170–182 GHz 185–190 GHz
Optical frequency	0.8–0.9 micron 1.06 micron 0.532 micron 10.6 micron

frequency. Table 1.2 lists the main frequency ranges used by the inter-satellite link in detail.

The selection of the working frequency of the inter-satellite link can be considered from the following aspects [17]:

1) It must meet the planning requirements of the International Radio Consultative Committee (CCIR-International Radio Consultative Committee) for the frequency of satellite inter-satellite links. UHF, VHF, Ka, S, and L frequency bands can be used for inter-satellite link communication.

   In the navigation satellite system, since the satellite–ground link has already used the S and L frequency bands, considering the electromagnetic compatibility of the satellite system, these two frequency bands are preferably no longer used by the satellite inter-satellite link, but mainly from the UHF frequency band, VHF band, and Ka band.
2) Refer to the inter-satellite link frequency bands selected and used by the US GPS BLOCK IIR and GPS BLOCK IIF. One of the major improvements of the third-generation GPS satellite BLOCK IIR in the United States is the addition of inter-satellite links to provide communication and ranging between satellites, and to provide precise ranging data to GPS satellites for autonomous navigation. The data exchange of the navigation data generation system and the exchange of the autonomous navigation state vector data are carried out among the satellites in the system. All GPS BLOCK IIR satellites are equipped with transponder data units working in the UHF frequency band for inter-satellite link communication and ranging functions. This frequency-hopping spread spectrum communication system belongs to the time division multiple access (TDMA) method. There is also the inter-satellite link of the fourth-generation satellite GPS BLOCK IIF of the United States working in the UHF frequency

band, which is very important for the selection of the frequency of the inter-satellite link.

3) Consider from the perspective of reducing the influence of inter-satellite Doppler frequency shift. Because the relative motion speed between satellites is very large, in general, satellite communication signals will have a certain Doppler frequency shift, and the difficulty of link acquisition will increase correspondingly with the increase of Doppler frequency shift. For radial velocity v_0, the Doppler shift between two satellites at the operating frequency f_0 of the inter-satellite link is $f = 2v_0 f_0/C$. Considering the characteristics of the satellite constellation studied in this paper, the relative motion speed between satellites is very large, and the relative position relationship is always changing to choose from.
4) From the perspective of satellite inter-satellite communication and ranging antenna. For communication and ranging between multiple satellites in the navigation satellite constellation system, shaped wide beam antennas and multi-beam phased array antennas can be used. If a shaped wide beam antenna is used, the antenna beam should be able to cover the azimuth angle of 0°–360°, and the elevation angle of 45°–21° or 68°–44°. Due to the limitation of the transmit power on the satellite and the installation space of the antenna, the gain of the antenna is generally relatively high. Although the size of the antenna can be reduced by increasing the operating frequency of the system, the propagation path loss will increase accordingly. When the distance between the inter-satellite links reaches 60,000 km, the accuracy of the inter-satellite ranging will be reduced. However, if a multi-beam phased array antenna is used, due to the complexity of the relative position and distance relationship between satellites, each beam of the antenna is required to have the ability to automatically capture and track, and the direction of the antenna beam should be controlled according to the orbit. The necessary periodic transfer and switching are carried out according to the law, so the system design is quite complicated.

1.4 Microwave Inter-Satellite Link

1.4.1 Frequency Selection

Looking at the development of inter-satellite links in major navigation constellations in the world today, it is not difficult to find that microwave inter-satellite links are the most important and most widely used space communication system in current navigation systems. The frequency band selection of the inter-satellite link affects the communication capacity and communication quality in the satellite navigation system. At present, the inter-satellite microwave communication technology is developing toward the high-frequency band. According to the frequency allocation table of the Radio Regulations published in 2008, the frequency bands allocated by the International Telecommunication Union (ITU) to inter-satellite research and operation signals are shown in Table 1.3.

Summarizing the signal frequency bands in Table 1.3, it is not difficult to find that more microwave frequency bands are allocated to inter-satellite link services, and the bandwidth of microwave frequency bands is also increasing [7, 9].

Table 1.4 lists the frequency band division of the inter-satellite links of domestic and foreign satellite systems in turn. Combining Tables 1.3 and 1.4, it can be seen that the main frequency bands currently used for inter-satellite links are UHF, S, Ka, Ku, and V.

In addition, it can be seen from the development trend of microwave inter-satellite links that the main development direction of inter-satellite links is high-frequency band, large channel capacity, and wide bandwidth. The advantages of using high-frequency band for inter-satellite communication in the microwave inter-satellite link of the global satellite navigation system are as follows:

1) Comply with the ITU space operation frequency usage specification, allow application, and be protected.

Table 1.3 Inter-satellite link frequency bands allocated by ITU.

Band name	Frequency range	Business scope
UHF band	400.15–401 MHz, 410–420 MHz	Limited to space operations, satellite earth exploration, space research
S-band	2.025–2.11 GHz, 2.20–2.29 GHz	Aviation radio frequency band, satellite downlink navigation frequency band
C-band	5.01–5.03 GHz	Inter-satellite use
Ka-band	22.55–23.55 GHz, 24.45–24.75 GHz, 25.25–27.5 GHz, 32.3–33 GHz	Inter-satellite link services in geostationary orbit only
V band	54.25–58.2 GHz, 59–71 GHz	Inter-satellite only, geostationary applications only
mm band	116–123 GHz, 130–134 GHz, 167–182 GHz, 185–190 GHz, 191.8–200 GHz	Inter-satellite use
Laser	0.8–0.9 μm, 1.06 μm, 0.532 μm, 9.6 μm	Limited to space research, not considered a protected security business

Table 1.4 Frequency bands of inter-satellite links between domestic and foreign satellite systems.

Satellite system	Frequency	Construction practice
The first and second generation of TDRS system	S / Ku sum S / Ku / Ka	1983–2002
Military star second generation	V	1994–2003
Iridium system	Ka	1997–2003
GPSIIR and GPSIIF	UHF	1999–2012
Advanced very high-frequency satellite system	V	After 2010
TDRS third generation	S / Ku / Ka	After 2012
Next generation iridium system	Ka	After 2014
GPSIII	Ka	After 2015
GLONASS-M	S	Under construction
GLONASS-K and GLONASS_KM/NG	Gekikou	Under development
GALILEO satellite GIOVE_A	Gekikou	In-orbit test
"Beidou" second-generation navigation satellite	Microwave	On-orbit test

2) The available bandwidth of the high-frequency band is wider, the bandwidth of the high-frequency band is about 100 times that of the low-frequency band, and the communication capacity is larger.
3) The antenna beam is narrow and the anti-interference is strong.
4) The wavelength of high-frequency radio is small, the size of the signal transmitting equipment is small, and it is easy to be equipped with satellites.
5) It is less affected by the plasma in space, and the data obtained by inter-satellite measurement are more accurate.
6) Radio signals in high-frequency bands are not easily interfered by the ground electromagnetic environment.

1.4.2 Microwave Inter-Satellite Link Data Transmission System

For the global satellite navigation system, the satellites in the navigation constellation are medium and high orbit satellites, the number of satellites is high, and the complexity and precision are high, which makes the inter-satellite link networking of the navigation system a complex network, which is generally low. Orbital communication satellite system networking cannot be compared. In the navigation system, the work done by each satellite is the same, so the satellites in the constellation differ in space except for their spatial positions, which are essentially equal.

Therefore, the inter-satellite communication network formed by the navigation satellites is a centerless satellite inter-network, which has a large number of peer nodes.

The characteristics of the inter-satellite link networking make the inter-satellite link have strong topology adaptability and flexible and convenient network access. In addition, different from the inter-satellite link of the low-orbit communication satellite system, the main task of the inter-satellite link of the navigation satellite system is the communication and measurement between satellites, enabling information exchange between satellites and between satellites and ground stations.

In a satellite network, how to allocate links between satellite nodes to achieve maximum inter-satellite resource sharing is a key issue in the study of the working system of inter-satellite links. The multiple access methods currently existing in the satellite system include code division multiple access (CDMA), frequency division multiple access (FDMA), TDMA, space division multiple access (SDMA), and so on. The characteristics of various multiple access modes are shown in Table 1.5.

Comparing the advantages and disadvantages of various multiple access modes in Table 1.5, referring to the successful inter-satellite networking methods of the GPS system and the iridium system, it can be found that compared with CDMA and FDMA, TDMA and SDMA are more suitable for satellite networking of

Table 1.5 Comparison of multiple access modes.

Multiple access mode	Main allocation of resources	Advantage	Disadvantage
Code division multiple access (CDMA)	Numbers	Easy to accept, good anti-interference performance, good privacy	Complex acquisition and large inter-symbol interference
Frequency division multiple access (FDMA)	Frequency band	Simple equipment, mature technology, flexible receiving and sending	Easy to produce intermodulation interference, not suitable for large-scale networking
Time division multiple access (TDMA)	Time slot	Single carrier operation, mature technology	The time synchronization mechanism is complex
Space division multiple access (SDMA)	Space	Improve frequency band utilization and increase system capacity	High requirements for antennas and complex system control technology

navigation systems. The TDMA mode works on a single frequency point, which makes the onboard frequency point matching process simple, the hardware design is simple, and the maintainability and interchangeability between onboard devices are good. TDMA technology has great advantages in inter-satellite link networking. The characteristics of the SDMA mode determine that it is more suitable for microwave inter-satellites with high frequency and spot beam antennas.

This is an important development trend of inter-satellite link technology in future satellite navigation systems.

1.5 Laser Inter-Satellite Link

1.5.1 Technical Characteristics of Laser Inter-Satellite Link

The laser inter-satellite link is a new method for realizing high-speed data exchange between aircraft by utilizing the laser wavelength, high brightness, and high collimation characteristics. It is a new method different from the current, widely used inter-satellite radio frequency communication that has several features.

1) High communication rate (from 100 Mbit/s to more than 10 Gbit/s) → fast transmission speed: a single channel can provide a data transmission rate of more than 10 Gbps, which is far greater than the current data transmission of hundreds of Mbps in microwave communication rate. It can reach hundreds of Gbps or more through wavelength division multiplexing.
2) The beam divergence angle is very small → strong anti-interception ability. Different from radio frequency communication, laser communication adopts point-to-point communication mode, so it has the characteristics of high confidentiality, strong anti-interference, and strong interception ability.
3) Small size, light weight, low power consumption → flexible. The beam divergence angle is much smaller than that of microwave communication, and the antenna gain of space optical communication is much larger than that of microwave communication.
4) Away from the electromagnetic spectrum → strong anti-interference ability.
5) Easy to encode (load quantum key) → good security performance.
6) Large information capacity (three to five orders of magnitude wider than the RF bandwidth) → convenient band selection.
7) Communication requirements → The channel is unobstructed and can be seen.

The laser inter-satellite link is the detection of extremely long distance and extremely weak signals. It uses an optical telescope as an antenna for signal transmission and reception, which puts forward high requirements for a number of

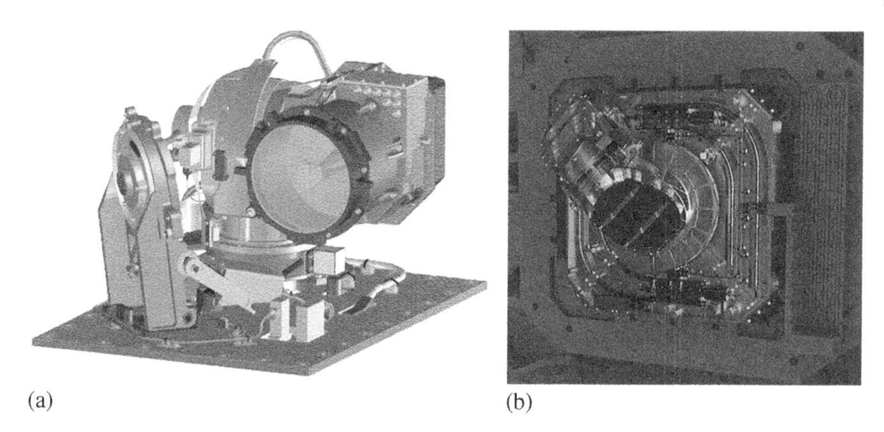

(a) (b)

Figure 1.4 Typical advanced space laser communication device. (a) US moon–Earth laser communication device; (b) German LCT laser communication device. *Source:* Deutsches Zentrum für Luft- und Raumfahrt.

optical performance and rotational characteristics of the device. The difficulty in development comes from the ultra-long distance, dynamic changes of links, and complex spatial environments. Establishing an inter-satellite communication link and keeping the link stable is the most critical technology. Satellites are always in high-speed motion, and several major steps are required to successfully establish a link and maintain stability, namely targeting, acquisition, and tracking. To realize the inter-satellite laser communication, the high sensitivity and high precision of the hardware are the foundation, while the speed and accuracy of the algorithm are guaranteed. For inter-satellite laser networking, issues such as multiple access, routing switching, and space network switching also need to be considered. For commercial laser communication constellations, it is also necessary to consider the coordinated development of performance and cost. It will not only be scientific issues, but also issues such as matching and compatibility with different industries as shown in Figure 1.4.

1.5.2 Future Requirements for Laser Inter-Satellite Links

The future demand for laser inter-satellite links may mainly focus on:

1) National deep space exploration and other major scientific projects

   The United States has achieved ultra-long-distance interstellar data transmission through the laser inter-satellite link, and the laser inter-satellite link can serve Chinese major scientific projects such as lunar exploration and fire detection.

2) Inter-satellite and satellite–Earth data interconnection
   Build a laser inter-satellite link constellation to achieve inter-satellite interconnection. It can be realized by deploying static orbit relay satellites, etc. In the occasions where satellite data needs to be quickly returned, such as disaster situation collection, emergency communication, enemy situation reconnaissance, satellite navigation, etc., the inter-satellite laser link can provide good real-time performance. For occasions that need to transmit large-capacity data, such as global mapping, weather detection, etc., the laser link can provide good stability.
3) A new generation of integrated laser payload
   The integrated payload that can realize various functions such as space target ranging, imaging, and communication may become the mainstream in the future.
4) Low-orbit Internet project
   The application of laser communication in Chinese low-orbit Internet project can improve the characteristics of inter-satellite interconnection, and is expected to improve the overall bandwidth and terminal performance.

1.5.3 Development Trend of Laser Inter-Satellite Links

1.5.3.1 The Development of Laser Communication Technology from Technical Verification to Engineering Application Stage

Laser communication technology is developing from the technology verification stage and the technology finalization stage to the engineering application stage. Many technical difficulties were gradually overcome. For example, fast and high-precision pointing, acquisition, tracking (PAT) technology, atmospheric turbulence effect suppression and compensation technology, narrow linewidth high-power laser emission technology, low-noise optical amplification technology, high-sensitivity DPSK/BPSK/QPSK optical receiving technology, etc. The overcoming of these technical difficulties has laid the foundation for the realization of interstellar laser communication.

1.5.3.2 The Communication Rate Develops from Low Code Rate to High Code Rate

The communication rate has been continuously improved, from the initial 2 Mbps to the current Gbps level, and the future planning has reached the order of tens of Gbps, gradually giving full play to the technical advantages of laser inter-satellite links. Early laser communication was mainly concentrated in the 800 nm light wave band. Various technologies in this band are relatively mature, the device performance is reliable, and the cost is low. However, its main disadvantage is the limited bandwidth of lasers and Si-APD detectors applied in this band. Therefore,

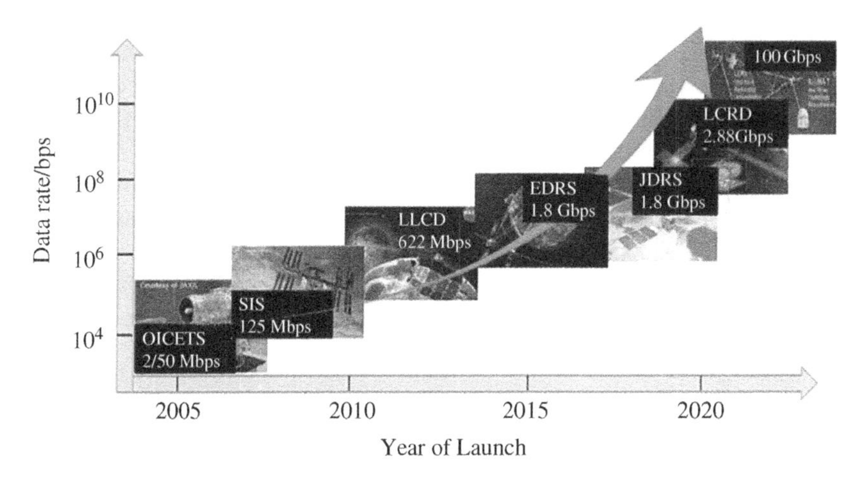

Figure 1.5 Schematic diagram of high-speed laser inter-satellite link.
Source: lilac 的老父亲.

the communication code rate is relatively low, generally less than 1 Gbps. At present, using the 1550 nm band for laser communication can make full use of the high bandwidth of the 1550 nm band laser transmitting and receiving components, and directly apply the mature technology of ground optical fiber to communication to improve the communication code rate as shown in Figure 1.5.

The main difficulties of high-speed laser inter-satellite links are key technologies such as high-speed optical transmission and high-sensitivity reception. The main technical approaches include high-order modulation technology (QPSK/DQPSK/M-QAM, etc.), optical multiplexing technology (wavelength division/time division/polarization/orbital angular momentum, etc.), high-sensitivity coherent receiving technology, etc.

1.5.3.3 Deep Space Will Become an Important Place for Laser Communication Applications

The main advantage of deep space laser communication is that it can realize the return of information from ultra-long-distance deep space exploration missions such as the moon, Mars, and Jupiter. The main difficulty is key technologies such as high-power optical emission and high-sensitivity reception. The main technical approaches include ultra-high-power optical emission technology, large-diameter optical antenna technology, and high-sensitivity single-photon detection technology.

With the continuous improvement of the return rate requirements of scientific missions, free-space optical communication has great potential to meet the high data rate links from deep space, and optical communication will be gradually applied in the deep space field in the future. Optical communications will enable the transmission speeds required for future scientific instruments. For example, the maximum transmission rate of the Mars Reconnaissance Orbiter (MRO) is 6 Mbps (the highest transmission rate for the current Mars exploration mission), which takes about 7.5 hours to empty the orbital recorder and 1.5 hours to transmit a high-resolution photo to Earth. The transmission rate of optical communication will be increased to 100 Mbps, and it only takes 26 minutes to clear the recorder, and the transmission speed of the same photo is only five minutes. Therefore, for future deep space scientific missions, the need to develop new high-speed transmission terminals is self-evident [16].

1.5.3.4 Combined Use of Laser Communication and Laser Ranging

Since laser ranging and laser communication have certain similarities in system composition, signal acquisition, and processing methods, they can be combined together to realize a functional whole to complete ranging and communication functions. The transformation plan of the SLR2000 satellite laser ranging station is proposed to combine laser ranging and laser communication, which is the original concept of the composite system. Since then, the European LISA has inverted the Earth's gravitational wave field through precise distance measurement, and achieved precise distance measurement at ultra-long distances with coherent laser heterodyne, which also has communication functions.

The main advantage is that the combination of communication and ranging enables a device to have multitasking capabilities, thereby reducing the requirements for volume and power consumption, and improving the cost-effectiveness of the system. The main difficulties are poor anti-interference ability and weak light energy for ranging. The main technical approach is to use common wavelength for ranging and communication, modulation dual system, pseudo-code and communication signal conversion technology.

1.5.3.5 Integration and Miniaturization of Laser Communication Terminals

In recent years, the United States, Europe, and Japan have been developing integrated, lightweight, and miniaturized laser communication terminals, which are mounted on small LEO satellites. The main advantages of integrated laser communication terminals are small size, light weight, low power consumption, good stability and low cost, and are usually carried on low-orbit small satellites. The main technical approaches are the lightweight and miniaturization of optical antennas and turntables, and the integration of communication transceivers.

In addition, for the commercial application of laser inter-satellite links, explore low-cost solutions to find a balance between ensuring product performance and reducing costs.

1.5.3.6 Networking of Laser Inter-Satellite Links

At present, the laser inter-satellite links in the world are all point-to-point, which seriously affects the communication relay, networking, and applications. Laser communication networking is an inevitable trend of future development. The main advantages of laser communication network are that the communication network is fast, real time, and wide area. The main difficulties are small beam divergence angle networking, dynamic topology access, and long delay. The main technical approaches are to break through the "one-to-many" laser communication technology, break through the "multi-standard compatible" laser communication technology, break through the all-optical relay technology, study dynamic routing to solve access problems, and seek a joint laser-microwave communication system.

References

1. Defense Information Systems Agency (DISA) (2012). Customer conference teleport program office (TPO).
2. National Aeronautics and Space Administration (2003). Space communications and navigation overview.
3. Younes, B. (2015). Space communications and navigation overview.
4. Tai, W., Wright, N., Prior, M. et al. (2012). NASA integrated space communications network.
5. Sodnik, Z., Smit, H., Sans, M. et al. (2014). Results form a Lunar laser communication experiment between NASA's LADEE satellite and ESA's optical ground station. *Proceedings of International Conference on Space Optical Systems and Applications (ICSOS) 2014, S2-1*, Kobe, Japan (7–9 May 2014).
6. Oaida, B.V., Abrahamson, M.J., Witoff, R.J. et al. (2013). OPALS: An optcial communications technology demonstration from the International Space Station.
7. Nappier, J.M. and Wilson, M.C. (2014). Long term performance metrics of the GD SDR on the SCaN testbed: the first year on the ISS. NASA/TM.
8. Kota, S.L. (2011). Hybrid/Integrated Networking for NGN Services. *2nd International Conference on Wireless Communication, Vehicular Technology, Information Theory and Aerospace & Electronic Systems Technology.*
9. Vanelli-Coralli, A., Corazza, G.E., Luglio, M. et al. (2009). The ISICOM architecture. *International Workshop on Satellite and Space Communications.*
10. Abraham, D.S. (2002). Identifying future mission drivers on the Deep Space Network. *Conference of Space Mission Operations and Ground Data Systems.* Houston, Texas, USA, 1–10.
11. Hurd, W.J. (2002). An introduction to very large arrays for the Deep Space Network. *Conference of Space Mission Operations and Ground Data Systems.* Houston, Texas, USA, 25–32.

12 Akyildiz, I.F., Akan, O.B., Chen, C. et al. (2004). Inter Pla Netary Internet state-of-the-art and research challenges. *IEEE Communications Magazine* 42 (7): 108–118.
13 Daniel, V. (2005). Deep Space Communication. *Space Physics C* 10 (8): 65–70.
14 Strategic, P. (2003). *National Aeronautics and Space Administration Publication*, 295–298. United States: Scientific and Technical Information.
15 Weber, W.J., Cesarone, R.J., Abraham, D.S. et al. (2006). Transforming the deep space network into the interplanetary network. *Acta Astronautica* 60 (8): 411–421.
16 Vachss, F., Norman, J., and Turner, C. (1999). Dual band mid-wave/long-wave IR source for atmospheric remote sensing. *Proceedings of SPIE* 3533: 174–179.
17 Fields, R., Fields, L.C., Wong, R. et al. (2009). NFIRE-to-TerraSAR-X lasercommunication results: satellite pointing, disturbance, and otherattributes consistent with successful performance. *Proceedings of SPIE* 7330: 73300Q-1-15.

2

Development History of Laser Inter-Satellite Link

2.1 Development Stage of Laser Inter-Satellite Link

Foreign laser inter-satellite link technology has achieved rapid development in recent years. The main research institutions are NASA JPL (Jet Propulsion Laboratory), NASA GSFC (Goddard Space Flight Center), MIT Lincoln Laboratory, and California Institute of Technology; Europe's ESA (European Space Agency), German Space Center, French Ministry of Defense Acquisition Agency; Japan's JAXA (Japan Space Agency), NICT (Japan's National Institute of Information and Communication Technology), etc. Figure 2.1 summarizes foreign laser inter-satellite links of major research institutes [1–3].

The development of laser inter-satellite link equipment can be mainly divided into three stages. The first stage is the principle exploration stage. From 1991 to 2000, as early as 1992, the Galileo probe observed two beams of ground-based lasers at a distance of 6 million kilometers. It successfully demonstrated the possibility of one-way signal detection from Earth to space. The first real space laser communication link was completed by Japan, which established the first successful space laser communication link between JAXA's ETS-VI GEO satellite and the 1.5 m NICT optical ground station (OGS) in Tokyo, Japan, in 1995; the rate reaches 1 Mbit/s.

The second stage was from 2001 to 2006, marked by direct detection technology. In 2001, Europe completed the first inter-satellite laser communication between the Artemis satellite and the SPOT-4 satellite. On the way, a laser altimeter was used to complete the two-way communication with the Earth, and the link distance was as high as 24 million kilometers. In 2006, Europe completed the world's first successful GEO satellite (ARTEMIS) and aircraft laser communication [4].

In the third stage, from 2011 to the present, the United States, Europe, Japan, and China have carried out laser inter-satellite link projects in the form of coherent detection. Compared with the first generation of direct detection laser communication technology with a rate of tens or hundreds of Mbps, the second generation

Laser Inter-Satellite Links Technology, First Edition. Jianjun Zhang and Jing Li.

Published 2023 by John Wiley & Sons, Inc.

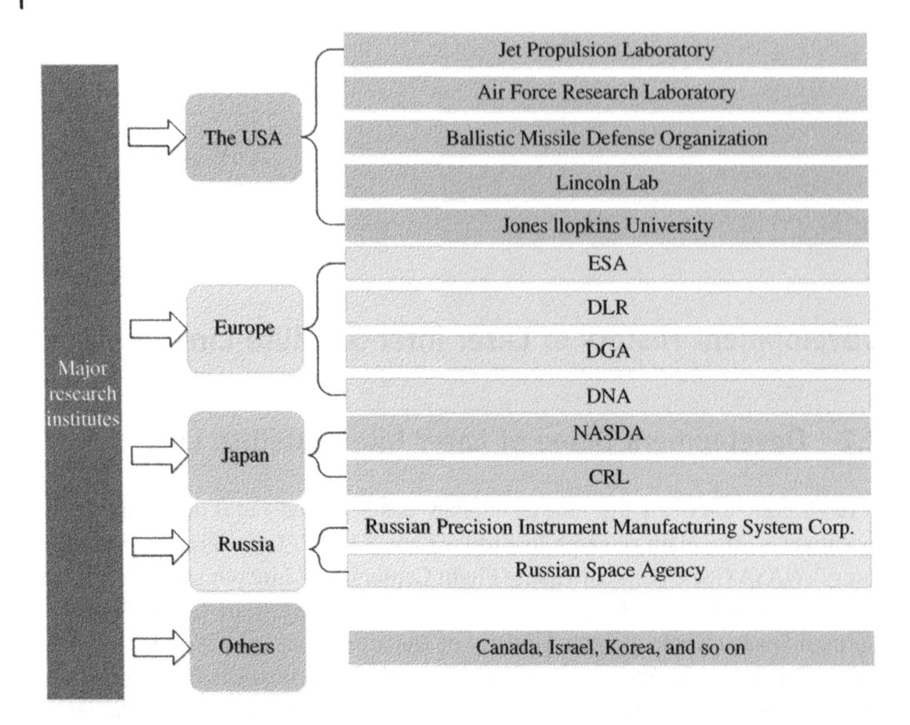

Figure 2.1 Major foreign research institutes for laser inter-satellite links.

speed of spatial coherent laser communication technology can reach several Gbps or even tens of Gbps, which is more practical. Laser inter-satellite links are gradually moving toward a practical stage, with more extensive applications. Deep space communication projects, relay satellite laser communication, high-speed inter-satellite communication, and quantum communication have been carried out one after another. Some commercial companies have also tried to apply laser communication to the space Internet [5]. Status of research on laser inter-satellite links in various countries is shown in Table 2.1.

2.2 Development Status of Laser Inter-Satellite Link Technology in Various Countries

2.2.1 United States

NASA in the United States has carried out research on laser inter-satellite link technology in the early days [6], such as "Laser Communication Demonstration System" (OCD), "Transformation Satellite Communication System" (TSAT),

Table 2.1 Status of research on laser inter-satellite links in various countries.

Country	The completed laser inter-satellite link experiment	Future plan
The USA	• 1995: GOLD(NASA JPL), GEO-GND, 0.8/0.5 μm, IMDD, 1 Mbps; • 2001: GeoLITE(NRO), GEO-GND • 2005: MESSENGER, Space-GND, 24 million kilometers • 2013: LRO(NASA GSFC), Lunar-GND, 1064 nm, PPM, 300 bps; • 2013: LLCD(NASA GSFC), Lunar-GND, 1550 nm, PPM, 622 Mbps; • 2014: OPALS(NASA JPL), ISS-GND, 1550 nm, IMDD, 30~50 Mbps; • 2016: Google X, near Space Exploration Balloon Room, 100 km, 155 Mbps; • 2019: LCRD(NASA GSFC), GEO-GND, 1550 nm, DPSK/PPM, 2.8 G/622 Mbps; • 2001: SILEX(ESA), GEO-LEO, GEO-GND, 0.8 μm, IMDD, 50 Mbps; • 2006: LOLA(ESA), GEO-Air, 0.8 μm, IMDD, 50 Mbps; • 2008: NFIRE(DLR), LEO-LEO, LEO-GND, 1064 nm, BPSK, 5.6 Gbps;	• 2021: ILLUMA-T(NASA GSFC), LEO-GEO-GND, 1550 nm, DPSK/PPM, 2.8 G/622 Mbps; • 2023: DSOC(NASA JPL), Mars-GND, 1060/1550 nm, PPM, 2 k/264 Mbps
Europe	• 2016: EDRS-A(ESA), GEO-LEO, GEO-GND, 1064 nm, BPSK, 1.8 Gbps • 2018: EDRS-C(ESA), GEO-LEO, GEO-GND, 1064 nm, BPSK, 1.8 Gbps; • 2018: OPTEL-μ(RUAG), LEO-GND, 1550 nm, IMDD, 2 Gbps;	• 2020: OPTEL-D(ESA), Deep space-GND, 1064/1550 nm, PPM, 192 kbps
Japan	• 1994: ETS-VI(NICT), GEO-GND, 0.8/0.5 μm, IMDD, 1 Mbps • 2006: OICETS(JAXA/NICT), LEO-GEO, LEO-GND, 0.8 μm, IMDD, 50 Mbps; • 2014: SOTA(NICT), LEO-GND, 980/1550 nm, IMDD, 10 Mbps • 2018: VSOTA(NICT), LEO-GND, 980/1550 nm, IMDD, 1k/100 kbps;	• 2021: HICALI(NICT), GEO-GND, 1550 nm, 10 Gbps

(Continued)

Table 2.1 (Continued)

Country	The completed laser inter-satellite link experiment	Future plan
	• 2019: JDRS(JAXA/ NICT), GEO-GND, 1550 nm, DPSK/ IMDD, 1.8 G/50 Mbps;	
China	• 2011: Ocean2(Harbin Institute of Technology), LEO-GND, 1550 nm, IMDD, 504 Mbps; • 2016: Mozi(Shanghai Institute of Optics and Mechanics), LEO-GND, 1550 nm, DPSK/PPM, 5.12 G/ 20 Mbps; • 2016: Tiangong2(Wuhan University), LEO-GND, 1550 nm, IMDD, 1.6 Gbps; • 2017: Practice13(Harbin Institute of Technology), GEO-GND, 1550 nm, IMDD, 2.5 Gbps • 2018: Beidou-3(Fifth Academy of Aerospace, Ninth Academy of Aerospace), IGSO-IGSO, IGSO-MEO, MEO-MEO, IGSO-GND, MEO-GND, 1550 nm, 1 Gbps • 2019: Practice20(Xi'an Branch), GEO-GND, 10 Gbps	

Lunar Laser Communication Demonstration (LLCD), and other projects. The later technological development has laid a good technical foundation. NASA is committed to promoting laser technology to become a common form of future space communication. In recent years, it has carried out a number of technology research and development projects: one is "Laser Communication Relay Demonstration and Verification" (LCRD) and the other is "Deep Space Optical Communication" (DSOC). It is expected to launch in 2023 as part of the Psyche mission, NASA's Discovery program. At that time, the "Psyche" spacecraft will carry the DSOC laser device to an asteroid composed of metallic elements to test the laser communication technology, which is said to be farther than the LCRD mission. The third is that the NASA Glenn Research Center team is developing the "Integrated Radio Frequency and Optical Communication" (IROC) concept, which plans to send a laser communication relay satellite to Mars orbit to receive data from long-distance spacecraft and relay the data to Earth as shown in Figure 2.2 [7–9].

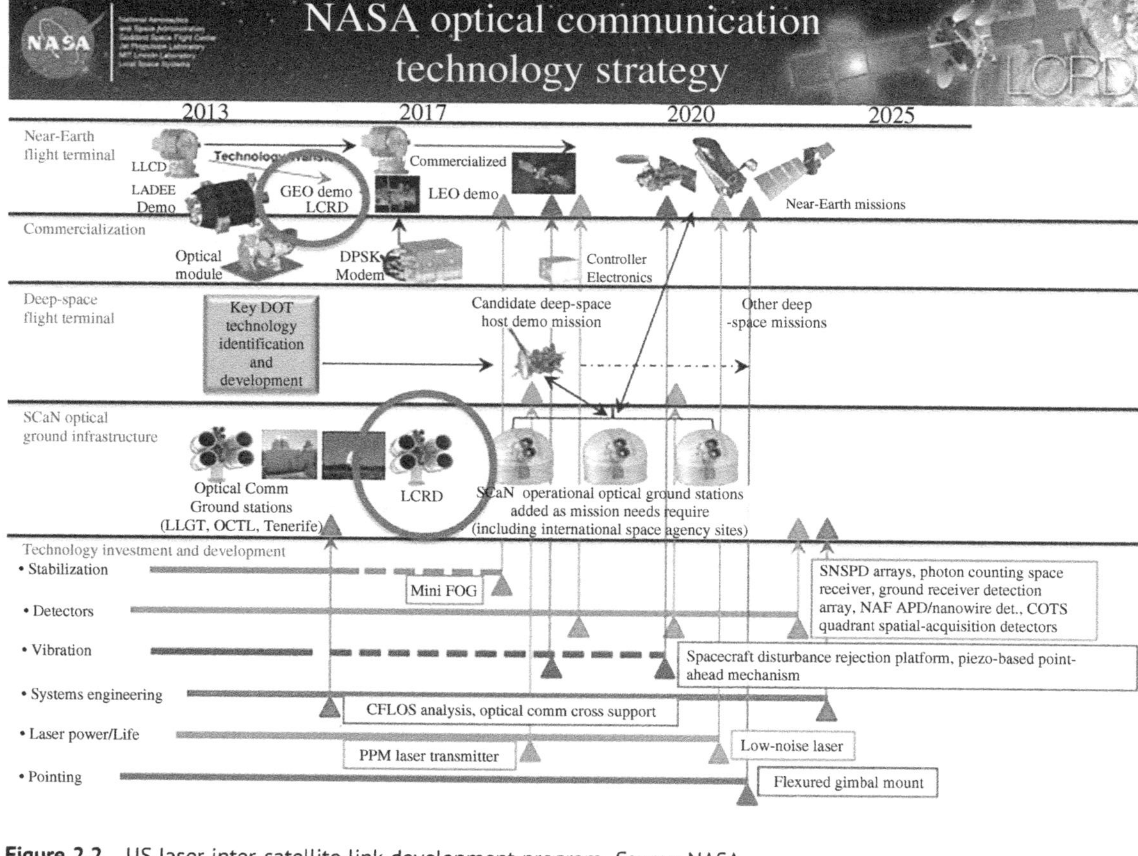

Figure 2.2 US laser inter-satellite link development program. *Source:* NASA.

2.2.1.1 Lunar Laser Communication Demonstration

The LLCD realized high-speed laser communication between the ground-based receiver and the lunar orbiter for the first time as shown in Figure 2.3. It is one of the most successful laser inter-satellite link projects in the United States [3, 8].

In 2013, the Lunar Atmosphere and Dust Environment Explorer (LADEE) carried the LLCD communication terminal and was launched into space. Under the conditions of thin cloudy weather, small horizon, and near sun, the communication terminal has successfully established a laser link. LLCD achieves data rates of up to 620 Mbit/s for the uplink and 20 Mbit/s for the downlink. In the upstream 20 Mbps test, the result of error-free transmission was achieved for the first time. This is the best result of the communication uplink between the moon and the Earth in human history. Compared with the current uplink data rate of microwave channels, which is only 1–2 kbps, the improvement is huge. Additionally, the payload exhibits laser ranging with an error of less than 200 ps bidirectional time-of-flight error providing sub-centimeter accuracy.

(1) Experimental goal
  - Verification of the laser communication technology under the Earth–Moon distance (about 380,000 km).
  - Real-time transmission of high-definition image data of lunar probes using laser channels.

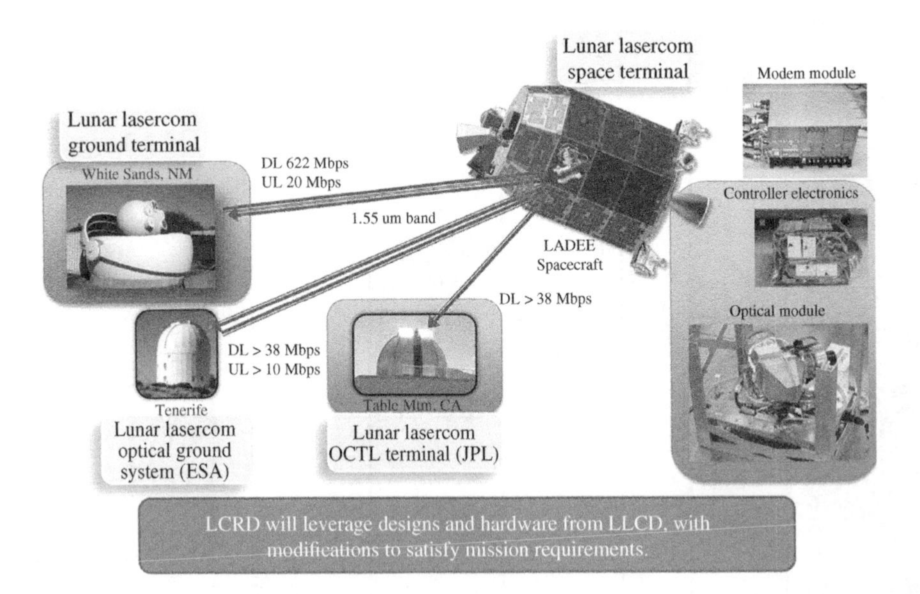

Figure 2.3 LLCD system composition. *Source:* NASA.

(2) Technical challenges
   - Long distance (380,000 km).
   - High data rate: Upstream 10–20 Mbps; Downstream: 40–622 Mbps.
   - Verification of laser penetration through thin cloud communication capability (Cloud Turbulence).
   - Verification of handover technology between ground stations.
   - Verification of Small Horizon Communication Capability (SEP).
   - Verification of the horizon communication capability.
   - Onboard large data storage and forwarding capability (current onboard forwarding channel is 39 Mbps).

The LLCD system consists of three main parts: the Lunar Laser Communication Space Terminal (LLST), the ground terminal (LLGT), and the operation center (LLOC). Its LLST optical module (OM) is a 10 cm Cassegrain telescope as shown in Figure 2.4 [9–11].

2.2.1.2 Relay Laser Communication Demonstration (LCRD) (GEO-Ground)

LCRD is a space high-speed optical communication demonstration project carried out by the United States; the purpose is to verify the laser inter-satellite link and network technology [12]. The space experimental satellite STPSat-6 was launched into space, mainly to carry out the two-way laser communication test between GEO–ground station, that is, the relay laser communication test of ground station–GEO–ground station. This mission will transform and use the ground station optical path in the LLCD mission. During the mission, the ground station in California, USA, will transmit a laser signal to a spaceborne laser communication terminal in geosynchronous orbit about 36,000 km away from the ground, and then the spaceborne laser communication terminal in geosynchronous orbit will relay the signal to another ground station. The project was jointly developed by NASA GSFC, NASA JPL, and MIT Lincoln Laboratory. There is no public report on the current operation in orbit [10, 11].

(1) Test target
   - Realize uplink 10 M to 20 Mbps, downlink 2.88 Gbps transmission capacity in GEO.
   - Verify key technologies of deep space laser communication.

(2) Technical challenges
   - Verification of high-orbit and low-Earth orbit communication terminal-to-ground terminal capture technology.
   - Onboard storage and forwarding capabilities of large amounts of data.
   - Verify modulation and demodulation and tracking technologies.
   - Verify key technologies such as DPSK technology and single photon counting.

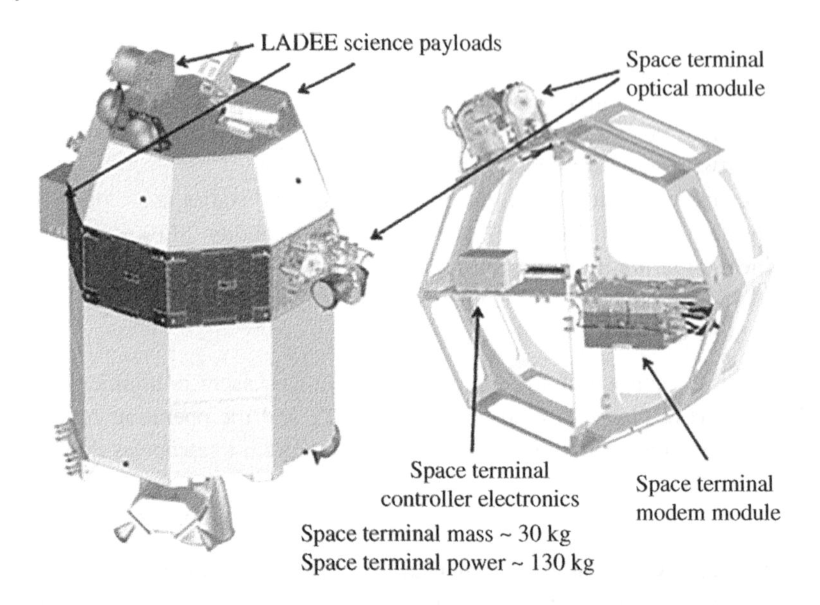

(a)

(b)

Figure 2.4 LADEE and LLST optical terminal module. (a) LADEE and Laser communication terminal; (b) LLST optical terminal module. *Source:* NASA.

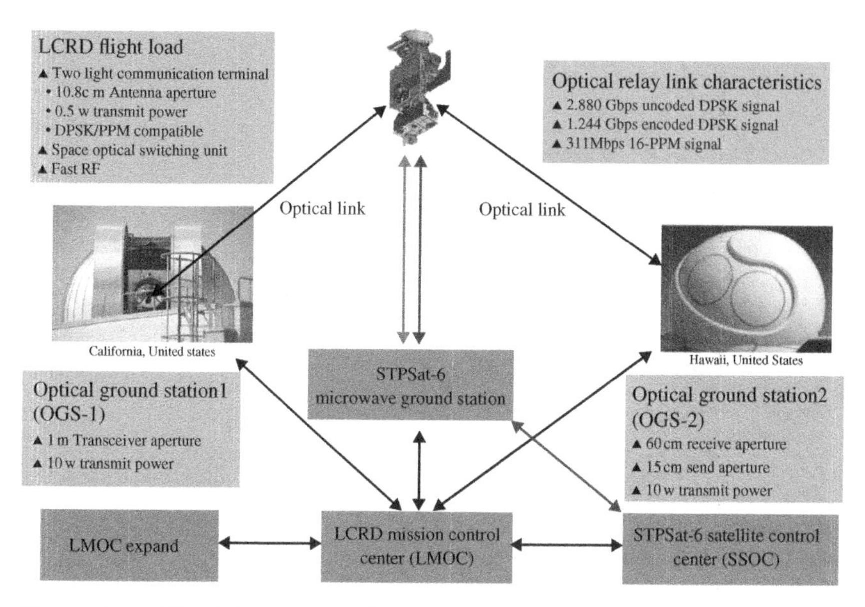

Figure 2.5 LCRD mission structure diagram. *Source:* NASA.

The LCRD mission structure diagram is shown Figure 2.5, which consists of two onboard laser communication terminals and two OGSs (one in Hawaii and one in California). The communication standard is compatible with DPSK and PPM, and the communication distance is 45,000 km [13–15].

The LCRD space payload includes two independent laser communication terminals (OST), which establish a laser communication link with the ground station or flight platform, respectively; the space exchange unit (SSU) manages the control, data routing, and remote control telemetry commands of the LCRD payload. Each OST consists of an OM, a modem, and an electrical control box (CE). The OM includes a Cassegrain telescope with an aperture of 108 mm and a two-axis gimbal. The transmitting and receiving optical signal is coupled to the telescope through a single-mode fiber. The modem supports PPM and DPSK signals, and the modulation rate is 2.88 Gbps. It can generate test data frames, and at the same time, it has a self-test function, which can complete calibration and nb loopback tests. The Electronic Control Box (CE) contains the Point, Capture, Track (PAT) software for the optical payload, receives feedback signals from the OM, generates control signals for the PAT software, and supports optical axis calibration and other functions. The Space Switch Unit (SSU) is the central controller of the payload. The SSU will receive and route user data, receive and process payload commands, and accumulate and transmit payload telemetry information. User data

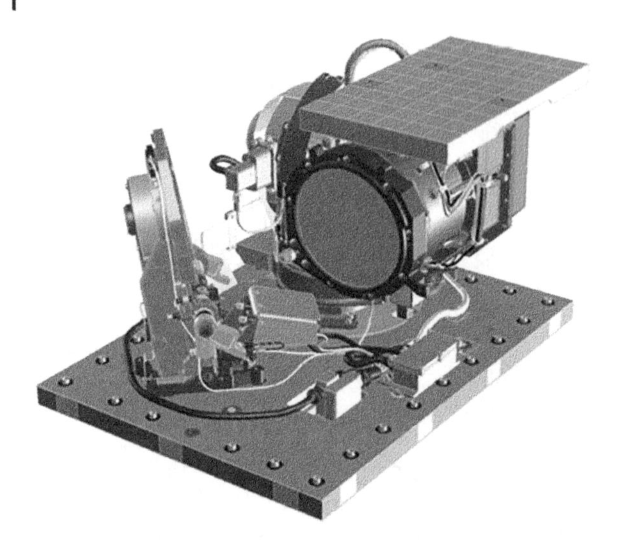

Figure 2.6 LCRD spaceborne optical terminal.

received from a space communication terminal can be routed to two destination ports at SSUnb (as shown in Figure 2.6) [6, 16, 17].

2.2.1.3 Integrated Laser Communication Terminal (ILLUMA-T)

This project is an expansion of the LCRD program. NASA plans to develop a low-cost near-Earth integrated ILLUMA-T terminal, which is expected to be launched to the International Space Station (ISS) in early 2021. The purpose is to establish a two-way communication link between GEO and LEO and complete the ISS–LCRD–ground station space networking [18]. The communication rate is 2.88 Gbps, the communication system is compatible with DPSK and PPM, the terminal mass is less than 30 kg, the power consumption is 100 W, and the cost of each terminal is expected to reach 5 million US dollars. The main objectives of the ILLUMA-T project include:

(a) Reduced size, weight, power consumption, and price of aerospace modems using integrated electronics/optoelectronics.
(b) Form an industrial chain of integrated LEO space modems (as shown in Figure 2.7).

2.2.1.4 Deep Space Optical Communication (DSOC) Project Terminal Reaches Level 6 Technology Maturity

In 2023, NASA plans to launch an exploratory metal satellite Psyche to run between Mars and Jupiter, and carry a laser communication terminal DSOC to conduct a series of deep space laser communication experiments with a

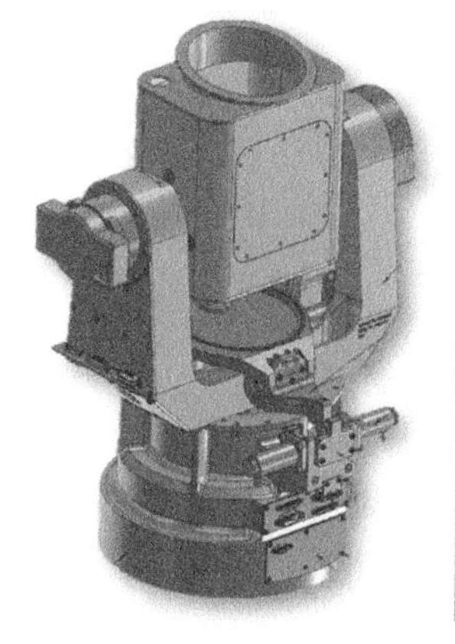

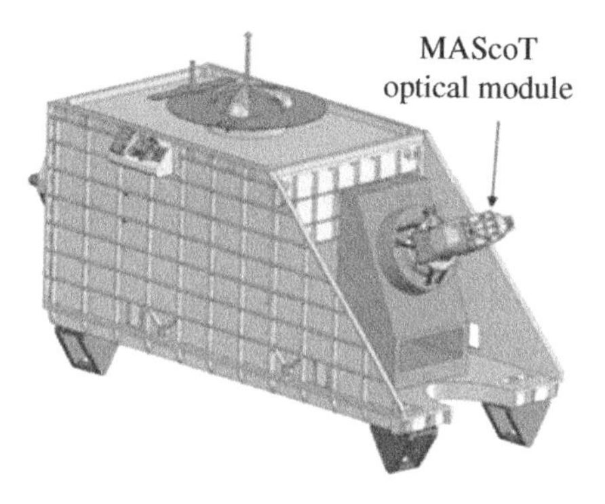

Figure 2.7 ILLUMA-T optical terminal and International Space Station deployment program. *Source:* NASA.

communication distance of 55 million kilometers; the project is expected to run to the working track in 2026. NASA says the DSOC has reached Technology Readiness Level (TRL) Level 6. TRL 6 means it is already a fully functional system prototype or representative model as shown in Figure 2.8.

The deep space spacecraft is equipped with a deep space laser communication terminal with a diameter of 22 cm, an emission wavelength of 1550 nm, and an average laser power of 4 W. The maximum communication rate can support serial pulse position modulation (SCPPM) of 267 Mbps.

The ground-based laser transmitter uses the 1 m diameter Hale Optical Telescope in Table Mountain, California. The wavelength of the laser signal is 1064 nm, and the maximum average power reaches 5 kW. The ground beacon light is used as the pointing reference of the deep space laser communication terminal, which can modulate 2 kbps LDPC encoded data. DSOC deep space exploration data transmission rate curve is shown in Figure 2.9.

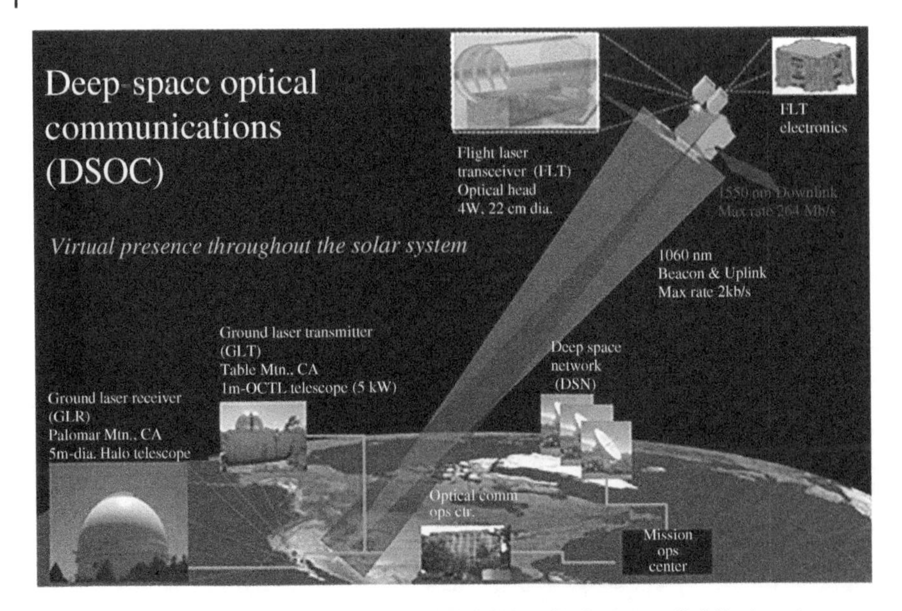

Figure 2.8 DSOC task diagram. *Source:* NASA/Wikimedia Commons/Public domain.

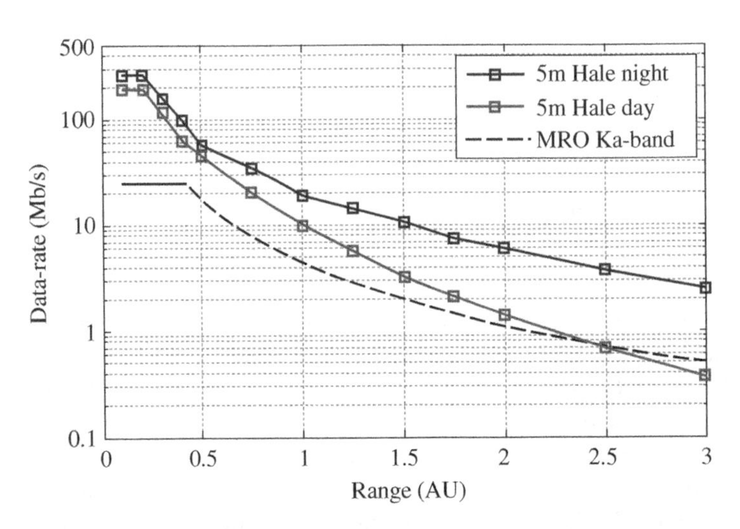

Figure 2.9 DSOC deep space exploration data transmission rate curve.

The ground-based laser receiver uses the 5-meter-diameter Hale Telescope in Palomar, California, to collect weak deep-space light signals on the downlink. Synchronization, demodulation, and decoding of received code words using an improved single-photon detector assembly with signal processing [19].

2.2.1.5 Ultra-Light and Small Communication Terminal (OSCD)

On 2 August 2018, the United States completed another milestone event in the field of spaceborne laser communication: the OSCD-B and OSCD-C terminals in the United States were each equipped with two 1.5U (1U refers to a standard unit with a volume of 10 cm × CubeSat AeroCube-7B and AeroCube-7C of 10 cm × 10 cm) completed satellite-to-ground laser communication based on the CubeSat microsatellite platform for the first time. The mass of the two communication terminals is only 360 g, and the downlink communication rate is 100 Mbps.

2.2.2 Europe

For the research of laser communication technology, Europe has successively formulated and implemented research plans related to satellite laser communication, resulting in the emergence of corresponding laser communication terminals. Representative research units and countries include ESA, German Space Center (Deutsches Zentrum für Luftund Raumfahrt, DLR), German Space Agency (GSA), France, Switzerland, etc. ESA's laser inter-satellite link network construction diagram is shown in Figure 2.10.

2.2.2.1 Semiconductor Laser Inter-Satellite Link Experiment

The Semiconductor Laser Inter-satellite Link Experiment (SILEX) carried out by ESA is the world's first inter-satellite laser communication link. The SILEX system consists of the GEO onboard terminal ARTEMIS (ESA 2001) and the LEO onboard

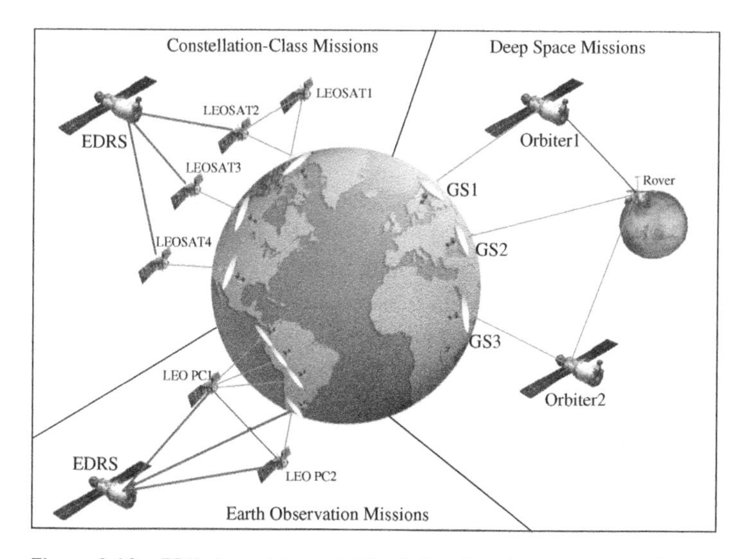

Figure 2.10 ESA's laser inter-satellite link network construction diagram.

terminal SPOT-4 (French 1998). As an important part of the SILEX program, the laser communication terminal carried by ARTEMIS successfully established a laser link with SPOT-4, which is 40,000 km away, and the data was transmitted from LEO to GEO at a rate of 50 Mb/s. The main technical challenges of SILEX are establishing laser communication terminal space applications, and direct modulation and detection of free laser inter-satellite links.

As a pioneer of inter-satellite link experiments, ARTEMIS has also conducted experiments with two-way communication with multiple other terminals. It includes the establishment of an inter-satellite link with the Japanese satellite terminal OICETS in 2005, and the successful realization of two-way laser communication. In 2006, LOLA, a mid-to-high-altitude unmanned aerial vehicle in France, realized the first star communication using the atmosphere as the medium. It shows that even with the influence of atmospheric turbulence, the incoherent communication method can realize the establishment of the star link and complete the star mission [4, 13].

2.2.2.2 European Data Relay Satellite System Project (EDRS)

After repeated tests and verification of the SILEX system and the mastery of mature technology, Europe began to deploy a new generation of laser communication data relay satellite system (European Data Relay Satellite System, EDRS) plan. The purpose of EDRS is to successfully establish a stable and reliable link between LEO satellites and GEO satellites, and to enable high-speed bidirectional transmission of information. Through the microwave communication between the relay satellite and the ground station, the inter-satellite high-speed optical communication and the ground communication are connected to form a hybrid communication network. The purpose is to use the inter-satellite high-speed laser communication to improve the information transmission capability of the entire communication network.

The EDRS consists of two GEO satellites, each carrying a laser communication payload (EDRS-A, EDRS-C, and EDRS-D) for interstellar information transmission. EDRS makes Europe no longer dependent on third-party base stations in other countries for high-speed spatial data transmission [6, 14, 15].

EDRS is jointly developed by ESA (research and development) and Airbus Defence and Space (responsible for the construction, launch, and operation of the system), and is the world's first independently operating commercial laser inter-satellite link system. The space segment of the EDRS Phase I system includes two geostationary orbit nodes, which are the EDRS-A data relay payload and the EDRS-C dedicated satellite configured with the data relay payload. EDRS-D is deployed in a static orbit over Asia, and is connected to other EDRS network nodes through laser communication to expand the service range.

In the several years that it has been put into daily operation services, the EDRS-A satellite has successfully completed tens of thousands of laser transmission connections, with a reliability of 99.8%, and can transmit 40 TB of data from remote sensing satellites, drones, and aircraft every day. ESA plans to expand into a global coverage system in 2020, form a space-based information network with laser data relay satellites and payloads as the backbone, and realize near real-time transmission of observation data from satellites and aerial platforms [16, 17, 19].

In addition to the Copernicus program that has already used the EDRS service, it will be applied to more other client terminals in the future, including the Columbus module of the ISS, the 2020 Pleiades Neo satellite (Airbus' high-resolution optical remote sensing satellite with a resolution of 0.3 m) will use this service.

The interstellar LCT is an upgraded version of the LCT carried by the German TerraSAR X satellite and the American NFIRE satellite. It compensates for the large space caused by long distances by increasing the laser transmission power, increasing the receiving optical aperture, and appropriately reducing the communication rate loss. Its main performance indicators are the communication distance is 45,000 km, the laser transmission power is 5 W, the receiving and transmitting antenna diameter is 135 mm, the communication rate is 1.8 Gbps, the communication standard BPSK, and the laser wavelength is 1064 nm (as shown in Figure 2.11).

2.2.2.3 Micro Laser Communication Terminal (OPTEL-μ)

In 2018, RUAG Space launched a miniature laser communication terminal named OPTEL-μ to LEO. The system consists of a low-orbit micro-space terminal and a ground terminal. The purpose of the project is to transmit the data generated on the LEO satellite to an OGS at a rate of 2.5 Gbps. The design of the micro space terminal follows the principles of light, small, stable, and multi-functional, serving various low-orbit small satellite platforms. The weight of the terminal is 8 kg, the volume is 8 l, and the power consumption is 45 W as shown in Figure 2.12.

The OPTEL-μ terminal design adopts a modular approach and consists of an optical head unit (OH, located on the nadir panel outside the spacecraft), a laser unit (LU, located on spacecraft nb), and an electronic unit (EU, located on spacecraft nb). It consists of a functional unit; OH, LU, and EU are interconnected by cables and optical fibers [12].

The OH performs the function of the PAT, ensuring the establishment and maintenance of the optical communication link during the passage of the satellite through the ground station, with a mass of 4.4 kg. The ground terminal adopts an optical telescope with a diameter of 0.6 m, and transmits an uplink optical signal of 1064 nm, 25 kbps 16-PPM modulation.

Figure 2.11 EDRS equipped with LCT-135 laser terminal developed by German Tesat company. *Source:* Matti Blume/Wikimedia commons.

2.2.3 Japan

Japan mainly adopts the method of international cooperation to carry out the research of laser inter-satellite link technology. The most successful development of communication terminals in Japan are the ETS-VI plan and the OICETS plan. In recent years, Japan has research and development plans for data relay, high-speed long-distance high-speed communication, and miniaturization.

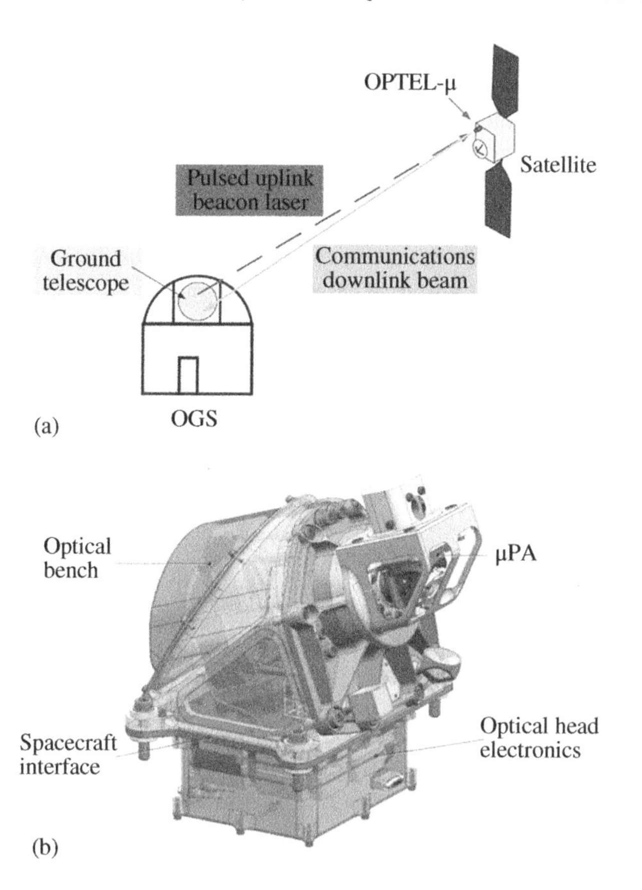

Figure 2.12 OPTEL-μ micro laser communication system. (a) System components; (b) OH optical termination module.

In the direction of miniaturization of laser communication terminals, Japan proposed the SOCRATES plan to verify the "Small Optical Communication Terminal" (SOTA) suitable for 50 kg-class small satellites. In addition, Japan is continuing to promote the "Laser Data Relay Satellite (JDRS)" plan and the "High-speed Communication for Advanced Laser Instruments (HICALI)" plan listed in the country's new "Space Basic Plan" in the field of laser inter-satellite links.

2.2.3.1 Japanese Data Relay Satellite

To meet the growing demand for high-speed data transmission, JAXA set out to develop a new optical data relay system. The system uses the data relay satellite JDRS to provide 1.8 Gbps data relay services through inter-satellite optical links and Ka-band feeder links. The mission of this project is not only to develop

GEO optical terminals, but also ground facilities and LEO optical terminals. The LEO optical terminal will be aboard JAXA's optical observation satellite "Advanced Optical Satellite." A demonstration of the optical data relay system will take place between the two satellites. The planned mission period is 10 years, during which JDRS will also support communications with JAXA's other LEO spacecraft.

2.2.3.2 High-Speed Communication of Advanced Laser Instruments

NICT has launched the HICALI project to promote research on next-generation laser inter-satellite link technology. The goal of the project is to achieve a 10 Gbps order, laser inter-satellite link from a geostationary satellite to an OGS with a communication wavelength of 1550 nm. The laser communication terminal will be launched into geosynchronous orbit on a high-throughput satellite (HST) in 2021. It will not only carry HICALI terminals, but also radio frequency (RF) terminals as shown in Figure 2.13.

The main objectives of the HICALI project are as follows:

(1) On-orbit verification of the first 10 Gbps GEO-to-OGS laser communication.
(2) On-orbit verification of novel optical modulation/demodulation methods.
(3) On-orbit verification of the reliability of new high-speed optical devices.

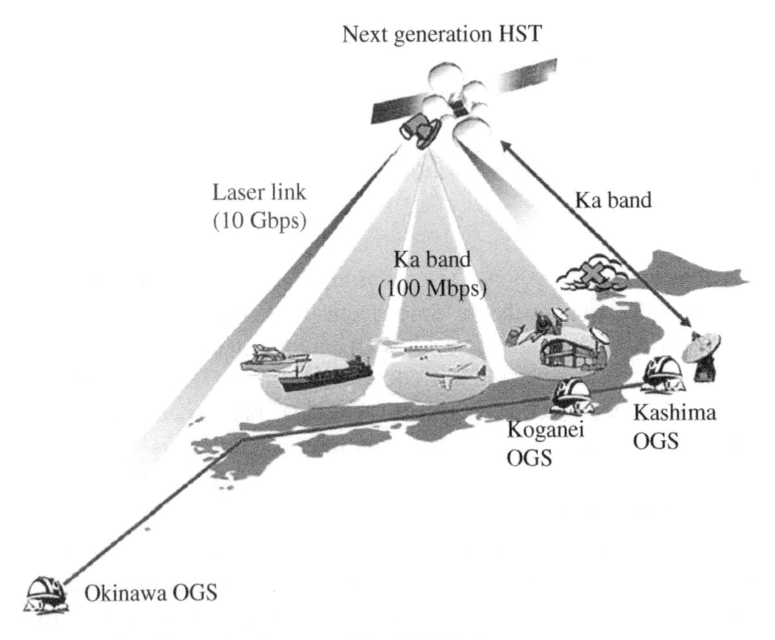

Figure 2.13 Schematic diagram of the HICALI plan.

(4) Acquisition of laser beam propagation data and accumulation of on-orbit experimental experience.

The HICALI project employs wavelength division multiplexing (WDM) technology devices, optical delay line interferometers, tunable laser assemblies (ITLAs), and high-speed digital processing devices.

2.2.3.3 Miniaturized Laser Communication Terminal (SOTA)

In 2014, through the "Space Optical Communication Research Advanced Technology Satellite" program (SOCRATES), Japan launched a "small optical communication terminal" equipped with a low-orbit small satellite SOCRATES into orbit, and verified the ground laser communication service of "Small Optical Communication Terminal" (SOTA) applicable to 50 kg small satellite. The total mass of the SOTA terminal is only 5.8 kg, the longest communication distance is 1000 km, and the downlink communication rate is 10 Mbps. SOTA can build a global optical communication network, so that the high-resolution image data collected by aircraft and satellites can pass through the laser inter-satellite link chain – down the road to the ground station.

2.3 Experience and Inspiration

With its bandwidth advantages, laser inter-satellite links are expected to become the main way of high-speed communication in space in the future. The United States, Europe, Japan, etc., have carried out comprehensive and in-depth research on various key technologies involved in the laser inter-satellite link system, and have developed a number of satellite laser communication terminals, and successfully completed a number of on-orbit tests. Space based laser communication network covering the world is being planned. Although Chinese laser inter-satellite link started late, during the 11th and 12th Five-Year Plan period, a large number of research contents of laser inter-satellite link have been arranged, a number of key technologies have been tackled, and a number of on-orbit demonstrations have been successfully conducted. Verify the project [17]. Compared with European and American countries, there is still a certain gap in inter-satellite communication and deep space laser communication.

After more than 40 years of development, foreign laser inter-satellite link technology has now entered the stage of engineering application and commercial application, and has opened up a new technology research upsurge. Facing the future needs of laser communication and drawing on foreign experience in the development of laser inter-satellite links, we have obtained the following inspirations:

2.3.1 Strengthen the Research on New Laser Inter-Satellite Links and Enhance the Innovation of Technology Research and Development

How to better improve the transmission capacity of the information network, the ability of free networking, and the availability of laser communication between the sky and the ground are the primary problems faced in the construction of the integrated information network between the sky and the ground. The emergence of new laser inter-satellite link technologies such as photon orbital angular momentum, phased array laser networking communication, and mode diversity reception provides new ideas and methods for the construction of space–Earth integrated information networks. China urgently needs to strengthen the research and development of new technologies for laser inter-satellite links and enhance innovation to meet the needs of scientific projects such as deep space exploration [12].

2.3.2 Strengthen the On-Orbit Verification of New Technologies and Improve the Engineering Level of New Technologies

After decades of technology accumulation abroad and multiple on-orbit test verifications, a solid foundation has been laid for the engineering application of laser inter-satellite links. In terms of laser inter-satellite link research, China should also gradually carry out various satellite laser communication on-orbit test verifications to rapidly improve the engineering level of satellite laser communication. At the same time, relying on the platform of China's space station, various types of laser communication links such as inter-satellite, star-ground, and starry sky can be tested and related technology verifications can be carried out to achieve leapfrog development and accumulate experience for the construction of future space–ground integrated laser communication networks [20].

2.3.3 Simplify the Product Spectrum and Promote the Construction of Product Pipelines

In the future, the main use scenarios of laser inter-satellite links that may be equipped in large quantities include inter-satellite communication between different orbits. In addition, small communication terminals will be expected to greatly increase the efficiency of microsatellites, especially CubeSats. Develop a few general-purpose inter-satellite laser communication products for main use scenarios, such as high-orbit Gbps-level high-speed communication terminals, medium-orbit miniaturized Gbps-level communication terminals, low-orbit inter-satellite trunk network Gbps-level high-speed communication terminals, low-orbit ultra-light small Mbps-class short-distance communication terminals, etc. Reduce product development pressure, increase production capacity, and delivery speed [18].

2.3.4 Respond to Commercial Product Demand and Reduce Product Cost

The speed of constellation deployment has become the most important competitiveness of satellite constellation networking, and various players have launched constellation deployment to start the competition. The possible trend in the later stage is that more Internet companies, communication companies, and equipment manufacturers will enter the field of satellite constellation networking to promote market demand and industrial chain upgrades. SpaceX and One Web's star chain plans are extremely large, posing unprecedented challenges to satellite production line construction and supplier system capabilities.

At present, the high cost of laser inter-satellite link equipment greatly limits its commercial development trend. The use of intelligent tools to carry out the construction of automated production lines is an important means to reduce the cost of satellite development. On the premise of ensuring core functions, the development process and testing process are simplified, and a large number of industrial-grade components are used, so as to achieve high integration and rapid iteration of products, and find a balance between cost, performance, and reliability.

2.3.5 The Key Development Direction of Low-Orbit Laser Inter-Satellite Link Engineering Demonstration Work

A key question is how much (user/feeder/inter-satellite) link transmission rate needs to be enough? How should the three links be matched to maximize their respective roles. Based on the positioning and requirements of the low-orbit Internet project, determine key indicators such as communication delay and coverage, quantitatively calculate constellation capacity and inter-satellite data communication requirements, determine system capacity bottlenecks, and optimize networking forms and inter-satellite connectivity.

For large-scale inter-satellite networking and routing requirements, focus on the chain building model of the 4409 satellites in the first phase of Star link, including whether to establish a chain between hybrid constellations and how to establish a chain.

References

1 Ghassemlooy, Z. and Popoola, W.O. (2010). *Terrestrial Free-Space Optical Communications. Mobile and Wireless Communications Network Layer and Circuit Level Design*, 355–392. London, UK: InTech.

2 Roberts, W.T. (2017). Discovery deep space optical communications (DSOC) transceiver. In: *Free-Space Laser Communication and Atmospheric Propagation*

XXIX, vol. 10096, 100960V. San Francisco, CA, USA: International Society for Optics and Photonics.

3 Araki, K., Arimoto, Y., Shikatani, M. et al. (1995). Performance evaluation of laser communication equipment onboard the ETS-VI satellite. In: *Free-Space Laser Communication Technologies VIII*, vol. 2699, 52–59. San Jose, CA, USA: International Society for Optics and Photonics.

4 (2006). Space probe breaks laser record: a spacecraft has sent a laser signal to Earth from 24 million km (15 million miles) away in interplanetary space. *BBC News*, 6 January. http://news.bbc.co.uk/2/hi/science/nature/4587580.stm

5 Edwards, C., Jr., Stelzried, C., Deutsch, L. et al. (1998). NASA's deep space telecommunications roadmap.

6 Sodnik, Z., Furch, B., and Lutz, H. (2006). Free-space laser communication activities in Europe: SILEX and beyond. *LEOS 2006-19th Annual Meeting of the IEEE Lasers and Electro-Optics Society*, 78–79. IEEE.

7 Breidenthal, J. and Abraham, D. (2016). Design reference missions for deep-space optical communication. *The Interplanetary Network Progress Report*, 42, 205.

8 Messier, D. (2013). NASA Laser system sets record with data transmissions from moon. *Parabolic Arc*. http://www.parabolicarc.com/2013/10/22/nasa-laser-system-sets-record-data-transmissions-moon/

9 Sun, X.L., David, R.S., Evan, D.H. et al. (2013). Free space laser communication experiments from earth to the lunar reconnaissance orbiter in lunar orbit. *Optics Express* 21 (2): 1865–1871.

10 Cornwell, D.M. (2017). NASA's optical communications program for 2017 and beyond. *2017 IEEE International Conference on Space Optical Systems and Applications (ICSOS)*. IEEE, 10–14.

11 Israel, D.J., Edwards, B.L., and Staren, J.W. (2017). Laser communications relay demonstration (LCRD) update and the path towards optical relay operations. *2017 IEEE Aerospace Conference*. IEEE, 1–6.

12 (2016). Start of service for_Europe's SpaceDataHighway. *ESA*.

13 Robinson, B.S., Shih, T., Khatri, F.I. et al. (2018). Laser communications for human space exploration in cislunar space: ILLUMA-T and O2O. In: *Free-Space Laser Communication and Atmospheric Propagation XXX*, vol. 10524, 105240S. San Francisco, CA, USA: International Society for Optics and Photonics.

14 Greicius, T. (2017). *Psyche Overview*. NASA https://www.nasa.gov/mission_pages/psyche/overview/index.html.

15 (2017). *Deep Space Communications via Faraway Photons*. Leonard: NASA https://www.jpl.nasa.gov/news/news.php?feature=6967.

16 European SpaceDataHighway forges 20000 successful laser links. *ESA*, 2 April 2019. https://www.esa.int/Our_Activities/Telecommunications_Integrated_Applications/First_satellite_in_European_SpaceDataHighway_forges_20_000_successful_laser_links

17 Tesat Celebrates 10 Years of Laser Communication in Space. http://www.tesat.de/en/laser.

18 Yamakawa, S., Chishiki, Y., Sasaki, Y. et al. (2015). JAXA's optical data relay satellite programme. *2015 IEEE International Conference on Space Optical Systems and Applications(ICSOS)*, 1–3. IEEE.

19 Inside the World's First Space-Based Commercial Laser-Relay Service. aviationweek.com (accessed 24 February 2018).

20 Baister, G., Greger, R., Bacher, M. et al. (2017). OPTEL-μ LEO to ground laser communications terminal: flight design and status of the EQM development project. *International Conference on Space Optics 2016*. International Society for Optics and Photonics, 10562, 105622U.

3

Spacecraft Orbits and Application

3.1 Overview

Spacecraft flying in space, or orbiting the Earth, or navigating in interstellar space, all fly under the action of gravity and according to certain motion laws; the atmosphere is also quite different. This special space environment has a great impact on the spacecraft's on-orbit operation or the effective performance of its functions. Understanding and mastering these basic issues and objective laws is an important prerequisite for effective space activities [1–3].

In the research and analysis of spacecraft motion, in order to facilitate description and analysis, it is necessary to establish various coordinate systems first. For example, to describe the orbit of the Earth's satellite, it is more convenient to use the Earth-centered equatorial coordinate system with the Earth's center of mass as the origin. Space probes or planets use the heliocentric ecliptic coordinate system with the center of the sun as the origin. The cosmic velocity is the minimum velocity required to launch a spacecraft from the ground to the cosmic space. The meanings of the three cosmic velocities belong to the basic theory of aerospace, and are briefly explained in this chapter.

The orbit of the spacecraft is the trajectory of the center of mass of the spacecraft, including the launch orbit, the running orbit, and the return orbit. The motion law conforms to Kepler's three laws. This chapter describes the orbital elements of the spacecraft motion state and their transformations under two-body motion. On the basis of a brief introduction of the types of low-Earth space orbits, the system describes the processes of entering orbit, maneuvering, deorbiting, and returning. The two-body motion assumes that the Earth and the spacecraft are mass points, and the motion of the spacecraft is only affected by the gravity of the Earth. In fact, the spacecraft will also be affected by some other perturbing factors during the

Laser Inter-Satellite Links Technology, First Edition. Jianjun Zhang and Jing Li.

Published 2023 by John Wiley & Sons, Inc.

orbital operation. The basic law of orbital motion of deep space exploration is briefly discussed [4].

The influence of the space environment on the spacecraft is manifested as a comprehensive effect, that is, an environmental factor can have various effects on the spacecraft, and the state of the spacecraft is also affected by various environmental factors. Therefore, studying the elements of the space environment and its impact on space activities plays an important role in the development and operation of spacecraft. The chapter will describe the elements and laws of near Earth space environment and deep space environment and their impacts on space activities, focusing on the impacts of the sun and its activities, neutral atmosphere, plasma, high-energy charged particles, Earth's magnetic field, solar electromagnetic radiation, micrometeoroids and debris on space activities [5–7].

3.2 Kepler's Laws

Kepler's law, also known as the "law of planetary motion," refers to the law followed by planets revolving around the sun in space. After observation and analysis, three laws were summed up successively from 1609 to 1619. Kepler's laws also apply to describe the motion of a spacecraft around the Earth.

3.2.1 Kepler's First Law

Also known as the law of orbits: The orbits of all planets around the sun are ellipses, and the sun is at one focus of the elliptical orbit. For spacecraft:

1) The orbit of the spacecraft around the Earth is an ellipse.
2) The center of mass of the ellipse formed by the two celestial bodies, the spacecraft and the Earth, always coincides with an intersection. Since the mass of the Earth is huge compared to the mass of the spacecraft, the center of mass coincides with the center of the Earth. Therefore, the center of the Earth is always at the focus. The parameters involved in Kepler's first law are as follows:
   ① Eccentricity e: Determines the degree of deviation of the ellipse from the ideal circle.
   ② Semimajor axis a and semiminor axis b: The long axis of the ellipse is the line between the two farthest points on the ellipse, and the short axis is the line

between the two closest points on the ellipse, *where the eccentricity value e is:

$$e = \frac{\left(\sqrt{a^2 - b^2}\right)}{a} \tag{3.1}$$

3.2.2 Kepler's Second Law

Also known as the Law of Area: For any planet, the line connecting it to the sun sweeps out an equal area in equal time. For a spacecraft, the line connecting the spacecraft and the center of mass of the celestial body sweeps the same area on the orbital plane in equal time.

No matter where the spacecraft is in orbit, the same area is swept every day. Combined with Kepler's first law, the spacecraft moves in an elliptical orbit around the Earth, and each point of the spacecraft's orbit is different from the Earth. Therefore, the spacecraft must move faster when it is closer to Earth to sweep an equal area.

3.2.3 Kepler's Third Law

Also known as the law of cycles: The ratio of the cube of the semimajor axis of the orbits of all planets to the cube of the planet's orbital period is equal. For spacecraft, it means that the larger the semimajor axis of the ellipse of the spacecraft orbiting the celestial body, the longer the period of operation, and the smaller the orbital speed of the spacecraft.

This law can be expressed by the following formula:

$$\frac{p^2}{a^3} = K \tag{3.2}$$

Among them, p is the period; a is the semimajor axis of the ellipse; $K = GM/4\pi^2$, M is the mass of the central celestial body.

3.3 Two-Body Motion and Orbital Parameters

3.3.1 Two-Body Movement

The two-body problem mainly studies the law of motion of two particles when there is only gravitational action between them.

When studying the orbital motion of the spacecraft, the two-body motion is the theoretical basis, it only considers the effect of the Earth's gravity, and the trajectory of the spacecraft around the Earth is a Kepler elliptical orbit.

In the Earth-centered equatorial inertial coordinate system, according to Kepler's law, the motion equation of the spacecraft can be expressed as:

$$\left.\begin{aligned} \ddot{x} &= -\frac{\mu x}{r^3} \\ \ddot{y} &= -\frac{\mu y}{r^3} \\ \ddot{z} &= -\frac{\mu z}{r^3} \end{aligned}\right\} \tag{3.3}$$

Among them: x, y, z represent the coordinates of the spacecraft, respectively; r represents the distance between the center of the spacecraft and the center of the Earth; and $r = \sqrt{x^2 + y^2 + z^2}$. μ is the Earth's gravitational constant.

Solved by Eq. (3.3):

$$\left.\begin{aligned} x &= x(\sigma_i,\ t) \\ y &= y(\sigma_i,\ t) \\ z &= z(\sigma_i,\ t) \\ \dot{x} &= \dot{x}(\sigma_i,\ t) \\ \dot{y} &= \dot{y}(\sigma_i,\ t) \\ \dot{z} &= \dot{z}(\sigma_i,\ t) \end{aligned}\right\} \tag{3.4}$$

In the formula, σ_i ($i = 1, 2, 3, 4, 5, 6$) represents six integral functions. Equation (3.11) represents the functional relationship between the integral constant and the position and velocity of the spacecraft. It can be seen from the formula that when the position (x_0, y_0, z_0) and velocity ($\dot{x}_0$, $\dot{y}_0$, $\dot{z}_0$) of the spacecraft are known at a certain moment $t = t_0$, a unique integral constant σ_i can be determined; the commonly used orbital elements σ_i are: semimajor axis a, eccentricity e, orbital inclination i, ascending node right ascension Ω, the perigee angle ω, the true perigee angle f, after the integral constant is uniquely determined, the spacecraft position and velocity at any time can be obtained by formula (3.11).

Two-body motion is plane motion, and the motion equation of the spacecraft in the Earth-centered orbital coordinate system can be written as:

$$\left.\begin{aligned} \frac{d^2x_t}{dt^2} + \frac{\mu x_t}{r^3} &= 0 \\ \frac{d^2y_t}{dt^2} + \frac{\mu y_t}{r^3} &= 0 \end{aligned}\right\} \tag{3.5}$$

Among them, x_t and y_t represent the coordinates of the spacecraft in the Earth-centered orbit coordinate system.

For polar coordinate transformation, $x_t = r\cos f$, $y_t = r\sin f$ (f is the included angle between the axis $\vec{r}$ and the axis X, that is, the true anomaly angle), and into formula (3.5), we can get:

$$\left.\begin{aligned} \ddot{r} - r\dot{f}^2 &= -\frac{\mu}{r^2} \\ r\ddot{f} - 2\dot{r}\dot{f} &= 0 \end{aligned}\right\} \tag{3.6}$$

The second formula in formula (3.6) can be directly integrated to get:

$$r^2\dot{f} = h \tag{3.7}$$

Among them, h is the integral constant, which represents the angular momentum of the unit mass.

The transformation made by $1/r = u$ and taking f as the independent variable, we can get:

$$\frac{d^2u}{df^2} + u = \frac{\mu}{h^2} \tag{3.8}$$

The general solution of formula (3.8) is:

$$u = \frac{\mu(1 + e\cos f)}{h^2} \tag{3.9}$$

which is:

$$r = \frac{h^2}{\mu(1 + e\cos f)} \tag{3.10}$$

In the formula, e is the integral constant vector, which is located in the orbit plane, and the direction r is parallel to the minimum direction $\vec{r}$, which is called the eccentricity vector. Equation (3.10) is the orbital equation of the spacecraft. From analytical geometry, this equation is a conic curve equation, which reflects the relationship between the distance r from the center of the Earth and the true anomaly angle f.

3.3.2 Track Parameters

Orbital parameters, also known as the number of spacecraft orbits, are parameters that represent the shape, size, space position, and direction of the orbital plane and the instantaneous position of the spacecraft on the orbit. According to Kepler's three laws, the instantaneous position of a spacecraft in space can be determined by the six Kepler orbital elements. The related meanings of the six orbital elements (a, e, i, ω, Ω, f) are shown in Table 3.1.

Table 3.1 The significance of the number of orbital elements.

Number	Name	Description	Definition	Note
a	Semimajor axis	Track size	Half of the major axis of the ellipse	The orbital period depends on the orbital size
e	Eccentricity	Track shape	Ratio of half the focal distance to the semimajor axis	Closed track: $0 < e < 1$ Open track: $1 < e$
i	Inclination	The inclination of the track surface	The angle between the orbital plane and the equatorial plane, measured counterclockwise from the ascending node	Anterograde orbit:$i < 90°$ Polar orbit: $i = 90°$ Retrograde orbit: $90° < i$
Ω	Ascending node right ascension	Rotation of the orbital plane around the Earth	The angle from the vernal equinox to the ascending node, measured from west to east	$0° \leq \Omega \leq 360°$ when $i = 0°$ or $180°$, Ω is unsure
ω	Argument of perigee	Orientation of the track within the track plane	The angle from ascending node to perigee along the direction of spacecraft motion	$0° \leq \omega < 360°$ when $i = 0°$ or $180°$, and when $e = 0$, ω is unsure
f	Near point angle	Position of the spacecraft in orbit	The angle from perigee to the spacecraft position along the direction of spacecraft motion	$0° \leq f < 360°$ when $e = 0, f$ is unsure

1) Semimajor axis a
2) The semimajor axis a refers to half the distance between the perigee and apogee of the ellipse, and represents the size of the orbit. The semimajor axis is also equal to the average radius, so this is a very important measure, an indicator of the orbit, a parameter that determines the size of the orbit, as shown in Figure 3.1. S is the center of mass of the spacecraft, F_1 is the center of mass of the Earth, and F_1 and F_2 are the two foci of the ellipse. When the F_1 and F_2 are coincident, the elliptical orbit becomes a circular orbit. For a circular orbit, the semimajor axis is the radius of the circular orbit, where r is the distance between the center of mass of the spacecraft and the center of the Earth, and $2c$ is the distance between the two foci.
3) Eccentricity e

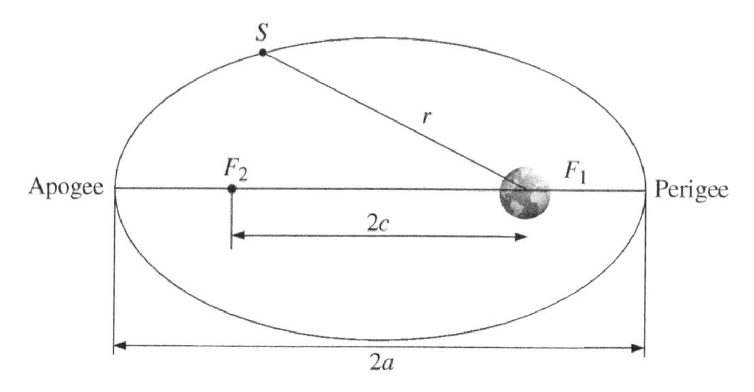

Figure 3.1 Schematic diagram of elliptical orbit.

Eccentricity determines the shape of the track. When $e = 0$, it is a circle; when $0 < e < 1$, it is an ellipse; when $e = 1$, it is a parabola; and when $e > 1$, it is a hyperbola. When $e \geq 1$, the spacecraft will escape from the gravity of the Earth and enter the solar system, becoming an artificial planet orbiting the sun, or going to other planets, as shown in Figure 3.2.

4) Orbit inclination

The orbital inclination is used to measure the angle at which the orbital surface is facing, indicating the degree of inclination of the orbital surface. The orbital

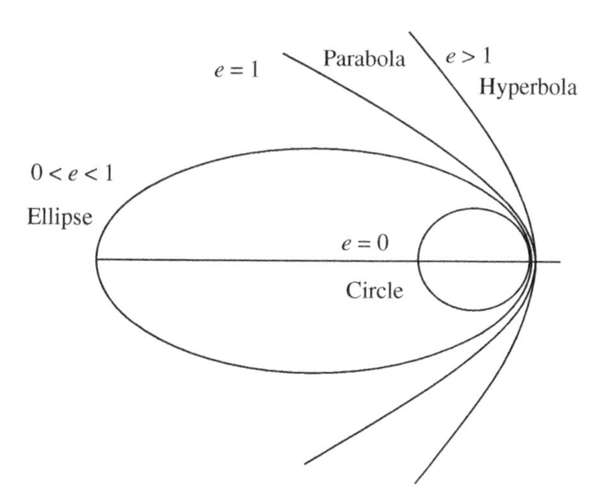

Figure 3.2 Orbit eccentricity.

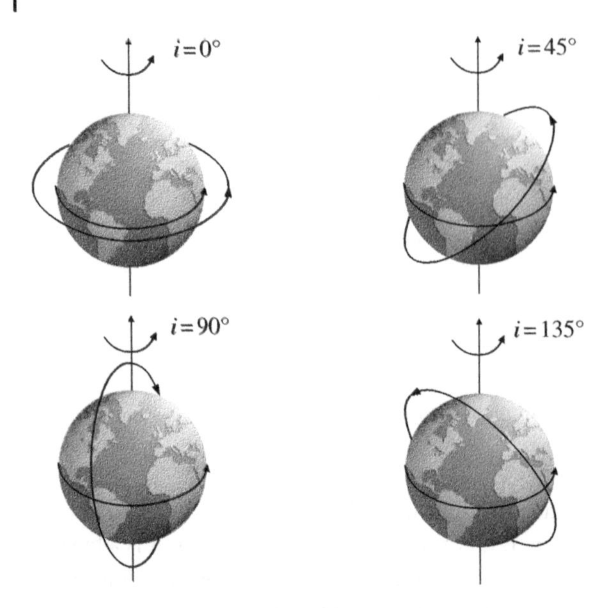

Figure 3.3 Schematic diagram of orbit inclination.

inclination is the angle between the orbital plane and the equatorial plane, as shown in Figure 3.3.

There are several orbits with different inclination angles: orbits with an inclination angle equal to 0° or 180°, whose orbital plane coincides with the equatorial plane, are called equatorial orbits. The inclination angle is equal to the inclination angle of 90°, and the spacecraft flies over the north and south poles of the Earth, which is called a polar orbit. If $i < 90°$, the orbit of the spacecraft is in the same direction as the rotation of the Earth (moving around the Earth from west to east), which is called a prograde orbit. If $i > 90°$, the orbit of the spacecraft is opposite to the rotation direction of the Earth (revolving around the Earth from east to west), which is called a retrograde orbit.

5) Ascending node right ascension Ω

The ascending node right ascension Ω is used to measure the rotation of the orbital plane around the Earth. It refers to the angle from west to east from the vernal equinox to the ascending node in the equatorial plane ($0 \leq \Omega \leq 360°$), as shown in Figure 3.4.

6) Argument of perigee ω

The argument of perigee ω is also called the angular distance of perigee, which represents the magnitude of the argument from the ascending node along the

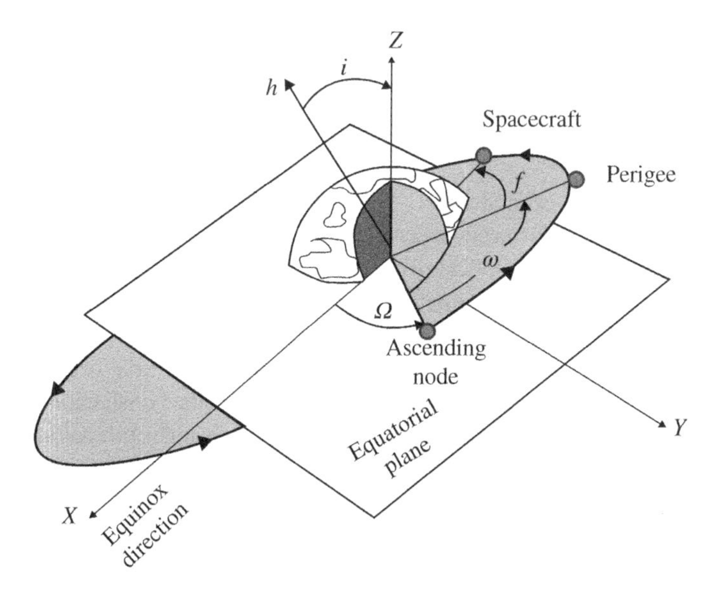

Figure 3.4 Ascension of the ascending node, argument of perigee and true perigee.

counterclockwise direction of the elliptical orbit to the perigee of the orbit. The argument of perigee represents the orientation of the orbit in the orbital plane. It refers to the angle from the ascending node to the perigee ($0 \leq \omega < 360°$) in the orbital plane according to the motion direction of the spacecraft, as shown in Figure 3.4.

7) True near point angle (f)

The true perigee angle is an angular quantity that describes the position of the spacecraft on the orbit at a specific point in time (epoch), which is the angle from perigee to the position of the spacecraft along the orbital plane along the direction of spacecraft motion ($0 \leq f < 360°$). The determination of the true anomaly is related to the position and time of the spacecraft, and it is the only element of the orbital element that changes with time, as shown in Figure 3.4.

3.4 Near-Earth Space Orbits and Applications

Usually, the space between 90 and 65,000 km from the ground is called near-Earth space, and its outer boundary is the range where the Earth's gravitational field can be ignored. Generally, the orbits of spacecraft operating in near-Earth space are collectively referred to as near-Earth space orbits.

3.4.1 Track Type

There are six orbital parameters in the low-Earth orbit of the spacecraft, and the orbits can be divided into various types according to the different values of the parameters.

1) Classification by spacecraft orbit shape:
   According to the shape of the orbit, spacecraft orbits can be divided into two categories: circular orbits and elliptical orbits.
2) Classification by spacecraft orbit inclination:
   According to the different inclination angles i of spacecraft orbits, spacecraft orbits are usually divided into three categories: $i = 0°$ called equatorial orbits, the orbital plane is on the equator, such as geosynchronous orbits. $i = 90°$ called a polar orbit, the orbital plane passes through the Earth's north and south poles. $0° < i < 90°$ is known as an inclined orbit, the orbital plane is inclined to the equatorial plane.
3) Classification by track height:
   Orbital altitude refers to the height of the spacecraft's orbit around the Earth from the surface of the Earth. According to the classification of orbit height, there are three types of orbits, that is, high orbits are orbits where the spacecraft's operating altitude is greater than 20,000 km. The medium and high orbit is the operating altitude of the spacecraft between 1000 and 20,000 km. Low-orbit satellites operate at an altitude of less than 1000 km.

3.4.2 Sub-Satellite Point Trajectory

The projection point of the spacecraft on the Earth's surface is generally called the sub-satellite point, and the change path of the sub-satellite point on the Earth's surface with time is called the sub-satellite point trajectory.

The surface of the Earth is a very complex surface, which is described by the elevation relative to a certain reference ellipsoid, so how to determine the exact position of the sub-satellite point? In orbit design, it is more convenient to approximate the Earth as a sphere, take the intersection of the line connecting the spacecraft and the center of the Earth with the spherical surface as the sub-satellite point, and use the latitude and longitude of this intersection to describe the position of the sub-satellite point and the trajectory of the sub-satellite point. It is the most direct method to reflect the law of spacecraft motion.

The longitude (represented by λ_s) and latitude (represented by φ_s) of the sub-satellite point of the spacecraft at any time t ($t > 0$) is determined by the following equations:

$$\lambda_s = \arctan(\cos i \cdot \tan(\omega + f)) + \Omega - \omega_e(t - t_0) - S_0 \quad (3.11)$$

$$\varphi_s = \arcsin(\sin i \cdot \sin(\omega + f)) \tag{3.12}$$

where λ_s is the geographic longitude of the sub-satellite point, φ_s is the geographic latitude of the sub-satellite point, i is the orbital inclination, ω is the orbital argument of perigee, f is true perigee angle at time t, Ω is the orbital ascending node right ascension, t_0 is the moment of passing the ascending node, ω_e is the angular velocity of the Earth's rotation, and S_0 is the sidereal time of Greenwich at time t_0, which is the right ascension.

The orbital inclination of polar orbit satellites generally ranges from 90° to 100°. Under this orbit, any lap of the satellite has an observation arc for the polar region. High-frequency observation of the polar regions can be achieved.

3.4.3 Several Commonly Used Tracks

The choice of spacecraft orbit is closely related to its mission function, different application tasks, such as remote sensing, communication, and navigation. Different track forms will be designed. Several commonly used tracks are as follows.

3.4.3.1 Sun-Synchronous Orbit

A sun-synchronous orbit is a spacecraft orbit in which the orbital plane rotates around the Earth's rotation axis in the same direction as the Earth's revolution, and the rotational angular velocity is equal to the average angular velocity of the Earth's revolution. The sun-synchronous orbit has the following characteristics: the local mean solar time and the annual variation of the illuminance at the sub-satellite point at the same latitude are the smallest. For the same latitude, the variation of the solar altitude angle is only related to the solar declination, and there are also only seasonal changes. Thanks to this characteristic of orbit, sun-synchronous orbit satellites image the same area with nearly identical lighting conditions. Ensure that the annual variation of the solar irradiation angle and solar energy received by the spacecraft is minimal, and the spacecraft operating in the sun-synchronous orbit can continuously and stably obtain solar energy. The terrestrial shadow time anniversary is almost unchanged. Since the precession speed of the spacecraft's orbital plane is almost synchronized with the apparent motion of the sun, the terrestrial shadow time change is also the smallest [8, 9].

The sun-synchronous orbit is realized by orbital perturbation. The influence of the Earth's aspherical perturbation J_2 term on the right ascension of the satellite's ascending node is used. The Earth's aspherical gravitational perturbation makes the orbital surface of the spacecraft rotate (precess) and is represented by $\dot{\Omega}$. If only the long-term effects of the harmonic term J_2 perturbation are considered, then:

$$\dot{\Omega} = -9.964 \cos i \left(1 - e^2\right)^{-2} \left(\frac{R}{a}\right)^{3.5} \tag{3.13}$$

The time that the center of the sun's apparent circle passes through the vernal equinox twice in succession is called the tropical year, also known as the solar year. Therefore, the angular velocity of the flat sun moving at the equator is:

$$\frac{\frac{360^\circ}{365.2422}}{d} = \frac{0.9856}{d} \tag{3.14}$$

where d stands for "every day."

A sun-synchronous orbit means an orbit in which the precession angular velocity of the spacecraft's orbital plane is equal to the angular velocity of the flat sun moving at the equator. According to this meaning, list the relational formula of sun-synchronous orbit:

$$-9.964 \left(\frac{R}{a}\right)^{3.5} \left(1 - e^2\right)^{-2} \cos i = \frac{0.9856^\circ}{d} \tag{3.15}$$

When the orbital eccentricity $e = 0$, that is, the spacecraft orbit is a circular orbit, Eq. (3.15) is simplified to:

$$-9.964 \left(\frac{R}{a}\right)^{3.5} \cos i = \frac{0.9856^\circ}{d} \tag{3.16}$$

According to the above formula, the relationship between the semimajor axis a of the sun-synchronous orbit and the orbital inclination angle i can be determined. When the inclination angle is greater than 90°, it precesses to the east, which is consistent with the direction of the Earth's revolution. Therefore, it is always a retrograde orbit.

Sun-synchronous orbits are not only related to the sun but also synchronized with the sun. Satellites operating in sun-synchronous orbits observe the same area at an almost fixed local time (mean solar time), which facilitates comparison of image data acquired at different times, so it has become the most widely used orbit for remote sensing satellites. This kind of orbit has high application value and is often used on spacecraft that are closely related to sunlight, such as resource satellites, meteorological satellites, and marine satellites.

3.4.3.2 Return to the Track

The return orbit refers to the orbit in which the trajectory of the spacecraft's subsatellite point changes again and again. The spacecraft running on the return orbit repeatedly flies over a certain area on the ground at a certain time interval, and can make periodic repeated observations of the same area to realize dynamic observation of the target area.

Assume that the angular velocity of the orbital plane of the spacecraft relative to the Earth is $\omega_e - \dot{\Omega}$, where ω_e is the angular velocity of the Earth's rotation in inertial space. Therefore, the time interval for one revolution of the orbital plane relative to the Earth is T_e, that is:

$$T_e = \frac{2\pi}{\omega_e - \dot{\Omega}} \tag{3.17}$$

Let the orbital period of the spacecraft be T_Ω (referring to the period of the intersection point), if there are both positive integers D and n satisfying:

$$nT_\Omega = DT_e \tag{3.18}$$

Then, after D days, the spacecraft runs exactly n circles, and its ground trajectory begins to repeat, such an orbit is the return orbit, and D is called the repetition period. Here $\dot{\Omega}$ is arbitrary, so the return orbit is not necessarily a sun-synchronous orbit. Only when $\dot{\Omega}$ is equal to the average angular velocity of the Earth's revolution around the sun, and the orbital period satisfies the formula (3.20), it is the sun-synchronous return orbit.

Remote sensing spacecraft usually have to consider (global) ground coverage, therefore, more use of return orbits, especially sun-synchronous return orbits. In this kind of orbit selection, the field of view of the remote sensor or the width at a certain height (a certain scale) should be fully considered. In this way, regular dynamic observations can be made on the same area on the ground with the same local time and observation conditions.

3.4.3.3 Geosynchronous Orbit

Geosynchronous orbit, also known as 24-hour orbit, is a prograde spacecraft orbit with the same orbital period as the Earth's rotation period, regardless of perturbation. A spacecraft running in a geosynchronous orbit passes over the same place at the same time every day and its stars. The lower point trajectory is a closed curve, and to people on the ground, the spacecraft appears in the same direction at the same time each day. Sometimes an orbit with a period equal to a fraction of the Earth's rotation period is called a geosynchronous orbit, and a circular geosynchronous orbit with an inclination of zero degrees is called a geostationary orbit.

Geostationary orbit spacecraft can cover more than 1/3 of the Earth's surface. If three spacecraft are evenly distributed in orbit, they can cover the whole world except near the north and south poles. The angular velocity of the spacecraft in geosynchronous orbit is the same as the angular velocity of the Earth's rotation.

The orbital period and the Earth's rotation period are 1 sidereal day, that is, 23 hours, 56 minutes, and 04 seconds. According to Kepler's law of motion, the orbital radius of the geosynchronous satellite is $R = 42255$ km and the orbital speed $v = 3.14$ km/s.

Geostationary satellites include synchronous, super-synchronous, and sub-synchronous. In $24/T_0 = n/D$ (n and D are the minimum number of laps and sidereal days required to realize the repetition of the sub-satellite point trajectory, respectively, and T_0 is the rotation period of the Earth), when n/D is 1, it is a geosynchronous spacecraft, and when it is $1/K$, it is geosynchronous. The spacecraft, when it is F, is a geosynchronous spacecraft, where K and M are both positive integers greater than 1. A geosynchronous orbit spacecraft can cover about 40% of the Earth's area. This orbit is commonly used by weather satellites, communication satellites, and broadcast satellites [10].

3.4.3.4 Freeze the Track

Frozen orbit refers to keeping the ground height of the satellite almost unchanged in the same area. The arch line of this orbit is stationary, that is, the semimajor axis of the orbit remains unchanged, and the shape of the frozen orbit remains unchanged, namely $\dot{e} = 0$ and $\dot{\omega} = 0$. Here is a brief description of the build conditions for frozen tracks.

The formula including the gravitational perturbation term of the Earth is:

$$\begin{cases} \dot{\omega} = -\dfrac{3nJ_2a_e^3}{2a^2(1-e^2)^2}\left(\dfrac{5}{2}\sin^2 i - 2\right) \times \left[1 + \dfrac{J_3a_e}{2J_2a(1-e^2)}\left(\dfrac{\sin^2 i - e\cos^2 i}{\sin i}\right)\dfrac{\sin\omega}{e}\right] \\ \dot{e} = \dfrac{3nJ_3a_e^2 \sin i}{4a^3(1-e^2)^2}\left(\dfrac{5}{2}\sin^2 i - 2\right)\cos\omega \end{cases} \tag{3.19}$$

In the above formula, if $(5/2)\sin^2 i - 2 = 0$, then $\dot{\omega} = 0$ and $\dot{e} = 0$ hold for any ω and e. These two specific dip angles are called critical angles. In order to solve the problem of high-latitude ground communication, a large elliptical orbit with $i = 63.43°$, $\omega = 270°$, a perigee height of about 1000 km, and a period of 12 hours can be taken. Because of $\omega = 270°$, the apogee altitude does not change, about 40,000 km; $\dot{\omega} = 0$, the apogee position is always at the north latitude 63.43°. A satellite can work for more than 10 hours in a day. Selecting such an orbit can make up for the defects that geostationary orbit satellites have a too low elevation angle to high latitude regions and cannot be used in polar regions.

For near-Earth Earth observation satellites, the critical inclination may not be appropriate. If we select:

$$\omega = 90°, \quad e = \frac{\sin i}{(\cos^2 i/\sin i - 2J_2a/(J_3a_e))} \tag{3.20}$$

When the value of e is small, the e^2 term can be ignored. At this time, $\dot{e} = 0$ and $\dot{\omega} = 0$ can also be obtained, indicating that the eccentricity and the argument of perigee of this orbit are "frozen" under the condition of short-period perturbation. Not only the orbit shape remain unchanged, but also the arch line no longer rotates, which can meet the higher level requirements of spacecraft remote sensing applications.

A frozen orbit is a stable orbit. The satellites of the Earth, Mars, and the moon all have frozen orbits due to the north–south asymmetry of the gravitational field. The frozen orbits of their satellites also have different properties due to differences in the gravitational fields of their host stars. The eccentricity of the frozen orbit of the Earth satellite is very small, which is very beneficial to satellite remote sensing.

3.4.4 Overlay

3.4.4.1 Coverage Area

As shown in Figure 3.5, the instantaneous height of satellite S at a certain moment is h, and the corresponding sub-satellite point is G. The tangent between the satellite and the Earth is called the geometric horizon of the satellite. The ground area surrounded by it is called the coverage area, which is the sum of the ground areas that the satellite may observe at that moment. The ground area outside the coverage area is called the coverage blind area. Let P be a point on the geometric horizon, called the horizontal point, then $\angle SOP = d$ is called the coverage angle of the satellite to the ground:

$$d = \arccos\left(\frac{R}{R+h}\right) \tag{3.21}$$

In addition, $\angle PSG = \alpha$ is the center angle of the satellite to the ground, twice the arc distance from P to G is called the coverage bandwidth S_W, and the area of the coverage area is A, then

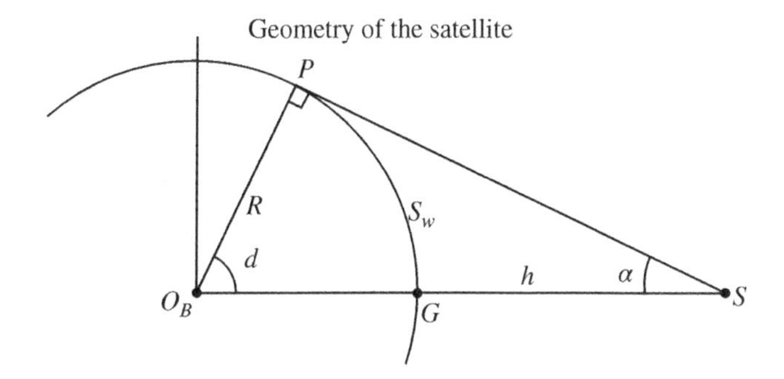

Figure 3.5 Schematic diagram of satellite coverage of the ground.

$$\alpha = 90° - d,\quad S_W = 2R{\cdot}d,\quad A = 2\pi R^2(1 - \cos d) = 4\pi R^2 \sin^2\frac{d}{2} \tag{3.22}$$

For geostationary orbit satellites, $d = 81.28°$, $\alpha = 8.63°$, $S_W = 18116.134$ km can be obtained, and A accounts for 42.54% of the world. Three geostationary satellites are placed at equal intervals on the equator to achieve global communication except for the blind spots near the poles.

3.4.4.2 Minimum Observation Angle

In the inner edge area of the largest coverage area, due to the influence of ground objects, the effect of satellite observation and communication is not good. In application, it is usually necessary to determine an effective coverage area, that is, it is stipulated that the angle between the line of sight SP and the horizontal plane cannot be smaller than a certain value δ, which is called the minimum observation angle, as shown in Figure 3.6, the corresponding coverage angle is recorded as d_δ, in ΔPSO, from the law of sine we can get

$$R\cos\delta = (R + h)\cos(d_\delta + \delta) \tag{3.23}$$

Then, the coverage d is correspondingly reduced to

$$d_\delta = \arccos\left(\frac{R\cos\delta}{R + h}\right) - \delta \tag{3.24}$$

Substituting d_δ for d in Eq. (3.24), the ground center angle, coverage bandwidth, and coverage area under the constraint of the minimum observation angle can be obtained.

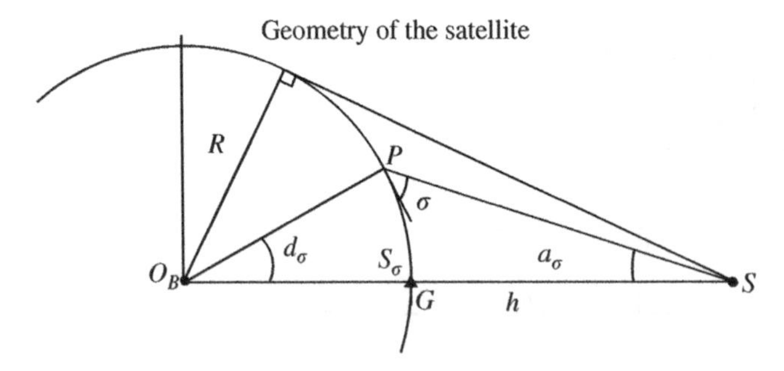

Figure 3.6 Schematic diagram of ground coverage under the constraint of minimum observation angle.

References

1 Clohessy, W.H. and Wihshire, R.S. (1960). Terminal guidance system for satellite rendezvous. *Journal of the Aerospace Science* 27 (9): 653–658.

2 Berreen, T.F. and Crisp, J.D.C. (1976). An exact and a new first-order solution for the relative trajectories of a probe ejected from a space station. *Celestial Mechanics* 13: 75–78.

3 Ryutaro, S. and Shigeru, M. A study of constellation for LEO satellite network. AIAA 2004-3236.

4 Heckman, N.E. (1986). Spline smoothing in a partial lineal model. *Journal of the Royal Statistical Society (Series B)* 48: 244–248.

5 Zeng, D. and Lin, D.Y. (2007). Maximum likelihood estimation in semiparametric regression models with censored data. *Journal of the Royal Statistical Society (Series B)* 69: 507–564.

6 Ma, Y., Chiou, J., and Wang, N. (2006). Efficient semi-parametric estimator for heteroscedastic partially linear models. *Biometrika* 931: 75–84.

7 Green, P.J. and Silverman, B.W. (1994). *Nonparametric Regression and Generalized Linear Models*, 17–21. London: Chapman and Hall.

8 Oaposchkin, E.M., von Braun, C., and Shamla, J. (2000). Space-based space surveillance with the spaced-based visible. *Journal of Guidance, Control and Dynamics* 23 (1): 148–152.

9 Stokes, G.H., von Braun, C., Sridharan, R. et al. (1998). The space based visible program. *Lincoln Laboratory Journal* 11 (2): 205–207.

10 Naka, R., Canavan, G.H., Clinton, R.A. et al. (1997). *Report on Space Surveillance, Asteroids and Comets, and Space Debris (Volume 3)*, 3–10. Washington: United States Air Force Scientific Advisory Board.

4

Basic Model of Constellation Inter-Satellite Link Networking

At present, there are many satellites in near-Earth space, and the types are complex, but they are basically independent, and it is difficult to achieve resource sharing. With the rapid development of the aerospace industry, the single-satellite working mode has gradually been unable to meet the needs of modern satellites to complete system tasks, and the network is the most effective way to complete the data interaction between nodes. Therefore, the network of satellite constellations is an inevitable trend in the development of the satellite industry. The satellite network composed of multiple satellites has good coverage, strong mobility, and good scalability at the same time. Especially, in the application of navigation constellations, it has a huge potential for action and so it has become one of the hot issues of international research. However, due to the complex structure of the satellite constellation and the time-space change of topology, the establishment of the basic model of its network has always been one of the basic problems studied by various researchers.

4.1 Application Requirements of Satellite Navigation Inter-Satellite Links

The inter-satellite link of navigation satellites is a dynamic wireless network with precise measurement and data transmission functions established between the satellites of the global navigation constellation system. Comprehensive analysis shows that the application requirements of inter-satellite links in satellite navigation systems are mainly reflected in several aspects such as constellation precise orbit determination and time synchronization, data communication, autonomous operation, and extended services [1].

Laser Inter-Satellite Links Technology, First Edition. Jianjun Zhang and Jing Li.

Published 2023 by John Wiley & Sons, Inc.

4.1.1 Constellation Precise Orbit Determination and Time Synchronization

The precise orbit determination and time synchronization of the navigation satellites are mainly completed by the ground master control station, the uploading station, and the monitoring station. The main task of the monitoring station is to carry out continuous observation and data collection of the navigation satellites through the receiver, and preprocess all the measurement data before sending it to the main control station. The main control station realizes the calculation of satellite messages by receiving and processing the data from all monitoring stations, and regularly updates the ephemeris through the registration station. Taking Beidou satellite navigation as an example, using the tracking of MEO satellites at domestic ground monitoring stations and Shanghai-based stations, the observable arc is less than half. The inability to monitor and forecast in time has led to a rapid decline in the URE index of satellites and a sharp drop in the positioning accuracy of the global system [2].

Therefore, for satellites with invisible arcs outside the country, it can only be solved by means of global station construction or inter-satellite links. The inter-satellite link can distribute the satellite ephemeris of the entire network to the entire constellation, and transmit the entire constellation observation data and key business data back to the main control station, improve the satellite orbit observation arc, and supplement the satellite–Earth observation data. The continuous inter-satellite measurement data provided by the inter-satellite link can be used for joint precise orbit determination and time synchronization between the satellite, the Earth, and the satellite. It can realize high-precision satellite orbit and clock difference calculation, realize the normal operation of global system navigation business, and ensure the realization of global system service performance indicators.

4.1.2 Data Communication

The data communication of the global satellite navigation system mainly includes global operation and control business data and global measurement and control business data. Through the inter-satellite link, the measurement and control capability and the coverage of business data communication can be improved to provide guarantee for the continuous monitoring and management of the global constellation. In the case that the measurement and control station does not have a global station, the navigation constellation MEO satellites have invisible arcs. Through the inter-satellite link forwarding, the global coverage of the navigation constellation in orbit can be achieved, and the uplink of the measurement and control instructions and the operation and control business data can be improved. The communication coverage capability of data transmission, telemetry downlink, and inter-satellite link scheduling management data. At the same time, through the inter-satellite link, the telemetry and remote control capability of the satellite in emergency situations can be improved. When there is no inter-satellite link, the satellite will not be able to transmit telemetry data back to the ground in real time in the invisible arc. If the satellite has an emergency failure outside the country, the ground will not be able to judge. Even if an emergency failure of the satellite is discovered in time, it is very likely that effective measures cannot be taken against the satellite because the satellite is in the invisible arc or because the time in the visible arc is too short. Therefore, the construction of inter-satellite links will largely meet the needs of telemetry and remote control of satellites in emergency situations [3–5].

4.1.3 Autonomous Operation

Navigation Constellation Autonomous Navigation means that when the constellation satellites are not supported by the ground system for a long time, through the inter-satellite two-way ranging, data exchange, and onboard processor filtering processing, the long-term forecast ephemeris of the satellites injected by the ground station is continuously corrected and timely and clock parameters autonomously generate navigation messages and maintain the basic configuration of the constellation to meet the needs of users for high-precision navigation and positioning applications. Navigation satellite of satellite system the inter-satellite link equipment to complete the inter-satellite ranging. The measured value includes the inter-satellite two-way measurement value, the transmission and reception time of the measurement signal, and the ranging correction information such as antenna correction; the inter-satellite link equipment is used to complete the injection. Navigation ephemeris provides satellite attitude information, routing information, inter-satellite signal distance and Doppler estimation, constellation satellite health, integrity and operation mode information, and other autonomous navigation information transmission. The autonomous operation of the navigation constellation improves the autonomous capability of the navigation constellation, reduces the dependence on the ground system, and can adapt to the needs of future satellite navigation system intelligence and navigation warfare. It has become the key direction of the development of major satellite navigation systems [4, 6].

4.1.4 Extended Service

The essence of the inter-satellite link is an integrated channel of measurement and data transmission. In some typical user scenarios, the inter-satellite link capability can be used to provide positioning of communication services for some specific extended user targets. High-speed near-space vehicles fly at high speed in near-Earth space. Due to the black barrier effect of penetrating the atmosphere, users cannot receive traditional L-band signals to obtain navigation services. When medium and high-orbit vehicles perform long-distance flight missions, there are insufficient ground tracking and measurement arcs, the difficulty of long-distance tracking, the difficulty in obtaining measurement and control data, etc., and space-based data service means are urgently needed [7, 8].

The satellite navigation inter-satellite link usually uses higher radio frequencies than the L-band, and the high-frequency band has the ability to penetrate the plasma black barrier; in addition, the navigation satellite inter-satellite link load also has the global coverage of the signal, and the measurement and data transmission capabilities are compatible. It can provide services for medium and high-orbit spacecraft navigation, deep space navigation, and high-speed near-space vehicles by using the redundant link resources of the navigation constellation inter-satellite links. Therefore, in addition to being an important support means for the operation of the global satellite navigation system, the inter-satellite link can also be used as an important means for strategic strikes and the construction of space-based systems [9, 10].

4.2 Basic Requirement Model of Inter-Satellite Link Network Application

4.2.1 Basic Configuration of Constellations

A satellite group consisting of several satellites distributed in a single orbit or multiple orbits as required is called a satellite constellation. The satellites of the satellite constellation work together to complete the same task under shared control, usually arranged in a special geometric shape according to a certain law. According to the classification of satellite orbit types, satellite constellations of a single orbit type can be mainly divided into polar/near-polar orbit constellations, geosynchronous orbit constellations, inclined circular orbit constellations, etc. There are also mixed orbit constellations composed of satellites in multiple orbits. Among them, there is a constellation called the Walker constellation in the inclined circular orbit constellation, which is a uniform symmetric constellation that consists of several orbital planes, and each orbital plane has the same number of satellites, which are evenly distributed in the orbital plane. All member satellites use circular orbits with the same height and inclination, and each orbital plane has the same angle [11].

Suppose the configuration of a Walker constellation is $N/P/F$, where N represents the number of satellites in the entire constellation, P represents the number of orbital planes, and F represents the phase factor of the Walker constellation, which is the ratio of the east orbital plane to the west orbital plane. The size of the front-facing phase is $(360°/N) \times F$, then in this Walker constellation, any satellite numbered i has the following relationship:

$$\begin{cases} \Omega_i = 360 \times \dfrac{P_i - 1}{P} \\ \mu_i = 360 \times P\dfrac{S_i - 1}{N} + 360 \times F\dfrac{P_i - 1}{N} \end{cases} \tag{4.1}$$

In the formula, Ω is the right ascension of the ascending node of the satellite, μ is the angular distance of the ascending node of the satellite, and P_i and S_i, respectively, represent the number of the orbital plane where the satellite is located and the number of the satellite in the orbital plane.

According to the classification of geometric types, Walker constellation can be divided into delta constellation, star constellation, rose constellation, among which Walker-delta constellation has been widely used because of its good coverage and simple structure. For example, the medium-orbit (MEO) navigation satellite constellation GLONASS, and the low-orbit (LEO) satellite constellation Iridium, all use the Walker-delta constellation.

In this paper, referring to the constellation design of Beidou navigation system, the object of analysis is the MEO Walker-δ constellation with an altitude of 20,000 km and an orbital inclination of 50°, and its orbital configuration is 24/3/1. In order to facilitate the analysis, calculation, and statistical data, the 24 MEO satellites are named as MEO11–MEO38, and the three orbital planes are named as orbital planes 1, 2, and 3, respectively, and they are placed in the high position of the number, and the low position of the number is the code of the satellite in the orbit; for example, MEO23 is the third satellite of the second orbital plane. In addition, on the basis of MEO satellites, three GEO and three IGSO high-orbit satellites have been added to enhance the regional coverage of the satellite constellation to make up for the disadvantages caused by insufficient ground stations. Among them, the longitudes of the ascending node of the three GEOs are 90°, 110°, and 130°, respectively, while the IGSO is three inclined geosynchronous orbit satellites with an inclination angle of 50°, longitude of the ascending node of 115°, and a phase interval of 120°. In this paper, the satellite tool kits (STK) software is used to simulate the structure of the satellite constellation. The simulation results are shown in Figure 4.1.

4.2.2 Inter-Satellite Transmission Network Based on STDMA

The networking of the inter-satellite link of the navigation constellation is a kind of complex satellite network. In this constellation, the satellites are medium and high-orbit satellites, with a large number, high complexity and high precision. This

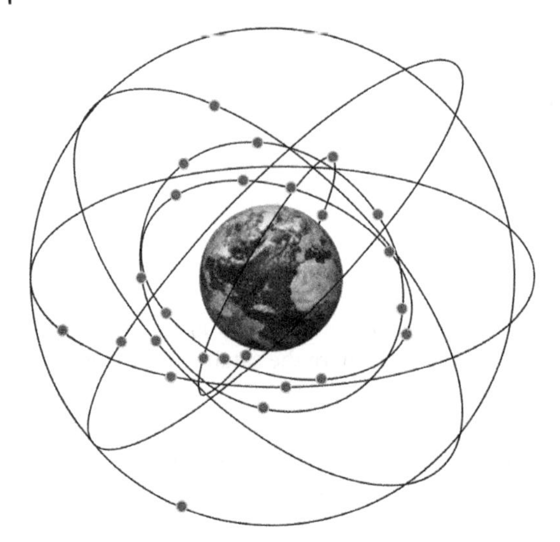

Figure 4.1 Simulation diagram of hybrid constellation structure.

is a general low-orbit small satellite. Networking is incomparable. Since the functions provided by the navigation satellites are consistent, each satellite of the navigation constellation is essentially equal, and the inter-satellite link network of the navigation constellation has the characteristics of no center, and so the inter-satellite link network of the navigation constellation can be regarded as a wireless network with a large number of peers [12].

The above characteristics of the navigation constellation inter-satellite link require its working system to have higher access flexibility and topology adaptability for the network construction of the inter-satellite link. In addition, the inter-satellite link network is also different from the general terrestrial wireless network. It is not intended to complete the intercommunication between nodes in the network. Its main function is to complete the connection between the ground station and the overseas satellites, so that the control instructions and measurement data can quickly reach the target node.

Considering the characteristics of these inter-satellite links, how to divide the node resources in the inter-satellite link network so as to cover more working areas and achieve resource sharing has become the most important issue in the design of the working system. At present, the main multiple access modes in satellite communication network systems are frequency division multiple access (FDMA), time division multiple access (TDMA), code division multiple access (CDMA), space division multiple access (SDMA), and so on. An overview of these multiple access modes is shown in Table 4.1.

From the above analysis of the advantages and disadvantages of various multiple access modes and referring to the networking strategies of GPS, Iridium, and other

Table 4.1 Comparison of basic multiple access modes.

Multiple access mode	Main allocation of resources	Advantages	Disadvantages
Frequency division multiple access (FDMA)	Frequency band	Mature technology, simple equipment, flexible sending	Easy to produce intermodulation interference, not suitable for large-scale networking
Time division multiple access (TDMA)	Time slot	Single-carrier operation, working in the saturation area, flexible networking	High time synchronization is required, and the synchronization mechanism is complex
Code division multiple access (CDMA)	Numbers	Easy to receive, strong anti-interference ability, concealed	Inter-symbol interference is large, and acquisition is more complicated
Space division multiple access (SDMA)	Space	Can increase the frequency band utilization, effectively increase the system capacity	System control is complex

systems, it is found that TDMA and SDMA are more suitable as the access modes of inter-satellite link networks; especially, TDMA access mode uses a single frequency point, and does not require complex and inflexible frequency pairing design. There is natural interchangeability between different inter-satellite link load devices, which is an important basis for realizing a centerless flat network. The complexity of the design and implementation of the inter-satellite link device is reduced, and it has a particularly good degree of matching with the inter-satellite link network [12, 13].

The access mode adopted in this paper is the time division space division multiple access mode (STDMA) which combines TDMA and SDMA. STDMA not only uses the space division mode to use directional antennas to form narrow beam switching directions to realize the multiplexing of the entire airspace, which improves the link gain and anti-interference ability, but also uses the time division method to realize the measurement and communication with multiple satellites which improves the overall network performance. In this way, each satellite can effectively establish an inter-satellite link with the target satellite under its own topological state, and will not interfere with other inter-satellite links at the same time. At the same time, the agility of the narrow beam antenna also ensures that the inter-satellite link system of the STDMA inter-satellite link system has stronger flexibility and feasibility in supporting other applications.

Time slot is the basic time unit under TDMA, and it is also the most important node resource. Each satellite has its own different link-building objects in each time slot, and due to the existence of SDMA, in STDMA access mode, each satellite has one and only one link-building object in each time slot object, that is, each satellite has only one inter-satellite link at the same time. In a time slot, the topological state of the inter-satellite link will not change. Only when the time slot changes, the topological state of the inter-satellite link may change accordingly.

The sum of a number of fixed time slots is called a superframe. In STDMA mode, a time slot reflects one satellite link establishment and then a superframe reflects the sum of all the links contained in it, called One chain building cycle. The change of the link establishment object of the time slot in the superframe and the change of the rules between the superframes are called link establishment planning, also known as time slot division.

Figure 4.2 is a schematic diagram of the time slot of a certain satellite. Assuming that the size of the superframe is three time slots, it can be seen that the time slot change of the satellite is the same in the first two superframes, and when the first two superframes are reached. Three superframes, the time slot division of which has been changed, are as follows. In the first superframe, the satellite is connected to the satellites SA, SB, and SC successively. In the second superframe, it is exactly the same as the first superframe, and repeats the chain establishment with these three satellites. In the third superframe, the satellites connected to the three time slots, respectively, become satellites SB, SD, and SE.

In STDMA access mode, all satellite nodes build links according to the allocated timeslots in the superframe, and each node can only have one link in a timeslot at the same time, as shown in Figure 4.3. The source satellite establishes links with the other five satellites in five different time slots, and after five time slots, that is,

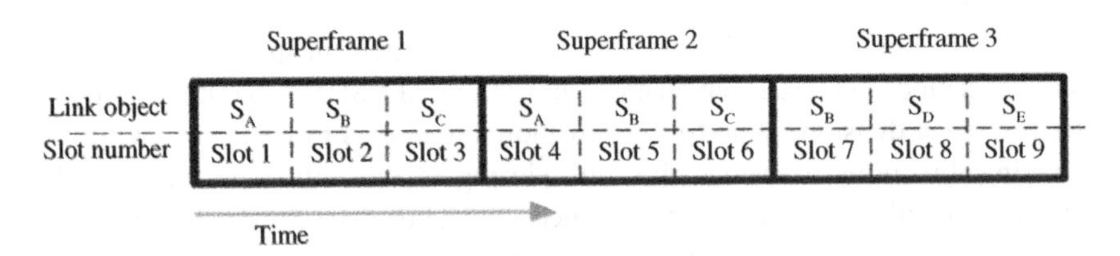

Figure 4.2 Time slot division change diagram of a satellite.

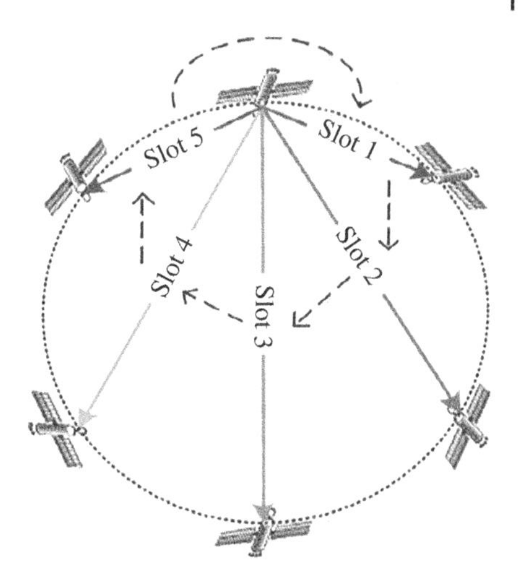

Figure 4.3 Schematic diagram of time slot access in STDMA mode.

one superframe polls to the first time slot and traverses it again. The direction of the dotted arrow in the figure reflects the time slot changes.

4.2.3 Antenna Solution

Since the STDMA multiple access mode is used in this paper, each satellite has at least one inter-satellite link at the same time, so it may be assumed that there is one and only one pair for the satellite on each satellite that constructs the inter-satellite link network antennas for inter-linking. In order to satisfy the space division condition of TDMA, it is required that the beams of the inter-satellite link will not interfere with each other. The traditional wide-beam reflector antenna has a large main lobe and weak side lobe suppression. When multiple inter-satellite links exist at the same time, mutual interference is likely to occur, so it is not suitable for use as a transmitting antenna for inter-satellite links. The narrow beam antenna not only has a small beam, concentrated energy, but also generates large power. In particular, its pointing is flexible and configurable, which can adapt to rapidly changing network topologies and realize the establishment of inter-satellite links in a short time [14].

In order to adapt to the STDMA access multiple access mode, the inter-satellite link antenna used in this paper is a narrow beam scannable antenna. The narrow beam scannable antenna can scan any area within its own beam range, and the switching rate is fast and the beam width is narrow. As shown in Figure 4.4, the beam scanning range of the narrow beam antenna used in this paper is set to α and the antenna. The normal direction always points to the center of the Earth.

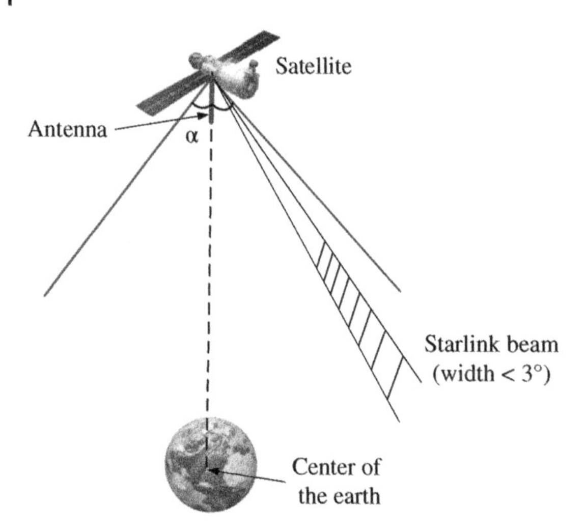

Figure 4.4 Schematic diagram of narrow beam antenna for inter-satellite link.

4.2.4 Inter-Satellite Link Application Mode

Under the current background, the inter-satellite link mainly exists to make up for the shortage of the number of ground stations on the ground. When the number of ground stations is severely limited, the operation and navigation services that rely on ground support are bound to be severely limited. Therefore, the completion of autonomous operation and autonomous navigation services is the biggest function of the inter-satellite link. It requires individual ground stations to manage and control the entire constellation of satellites, especially those that cannot be directly observed, as well as the distribution of ephemeris, global integrity monitoring of constellations, etc.

Figure 4.5 is a schematic diagram of the network composition of the entire constellation inter-satellite link and satellite–ground link. Since the model designed in this paper is for the inter-satellite link, it may be assumed that the satellite-to-ground link uses additional antennas and frequency bands, and does not interfere with the inter-satellite link. The ground station can send data to domestic satellites at any time. The time slot of the link is limited. If there is data that needs to be sent to overseas satellites, it will be relayed and forwarded through the inter-satellite link.

The types of data that the navigation constellation needs to transmit through the inter-satellite link include satellite autonomous operation and navigation data, operation control information, management and control information, satellite ephemeris, etc. Table 4.2 reflects the classification of inter-satellite data, its timeliness requirements, and link broadcast types.

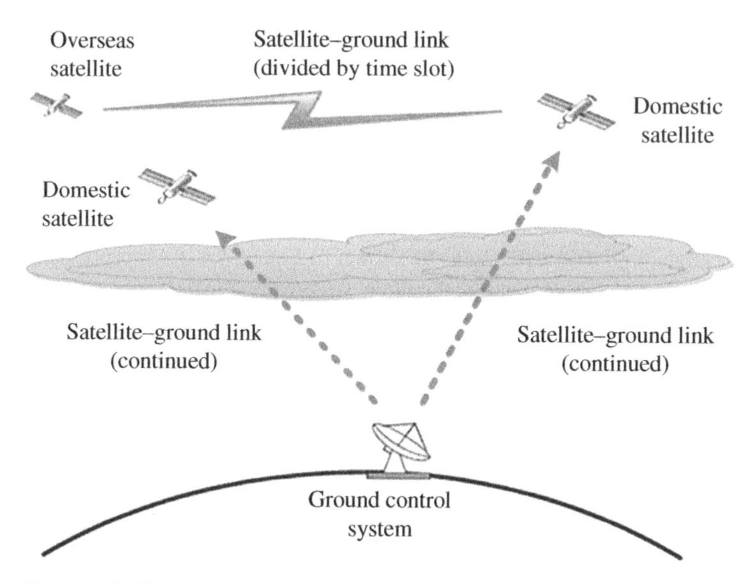

Figure 4.5 Schematic diagram of the composition of the constellation communication network.

Table 4.2 Inter-satellite link application data types.

Information category	Link type	Broadcast method	Timeliness requirements
Run control management	Ground station→satellite (→satellite)	Unicast/ Broadcast	High
Satellite ephemeris	Ground station→satellite (→satellite)	Unicast	Higher
Integrity data	(satellite→) satellite→ground station	Unicast	High
Inter-satellite ranging information	satellite→satellite→ (ground station)	Unicast	Direct transmission between satellites, lower back to ground station
Time synchronization information	Ground station→satellite (→satellite)	Unicast	High
Autonomous operation information	Ground station→satellite (→satellite)	Unicast	High
Ionospheric correction model	Ground station→satellite (→satellite)	Broadcast	Generally

4.3 Inter-Satellite Link Network Chain Topology Model

4.3.1 Analysis of Topological Attribute of Inter-Satellite Links

Topological attributes such as distance and angle between satellites directly affect whether the inter-satellite link can be established and the quality of communication after establishment. For satellites, the topological properties mainly include distance, pitch angle, and azimuth angle. As shown in Figure 4.6, RAB represents the inter-satellite distance, Φ represents the pitch angle of A star to B star, and η is the azimuth angle of A star to B star. This section will focus on analyzing the variation of visibility between satellites in the designed satellite constellation. Since the scannable area of the narrow beam antenna is a cone-shaped area pointing to the center of the Earth, its structure is symmetrical, and the change of the azimuth angle will not affect the visibility of the satellite. Therefore, this chapter mainly analyzes the distance and pitch angle attributes between satellite.

4.3.2 Inter-Satellite Visibility Analysis

Satellites can communicate with each other without any hindrance, which is called inter-satellite visibility. Inter-satellite visibility is the physical prerequisite for establishing inter-satellite links between satellites. There are many factors that affect inter-satellite visibility, among which, Earth occlusion and satellite antennas are the most critical factors affecting inter-satellite visibility [7, 12].

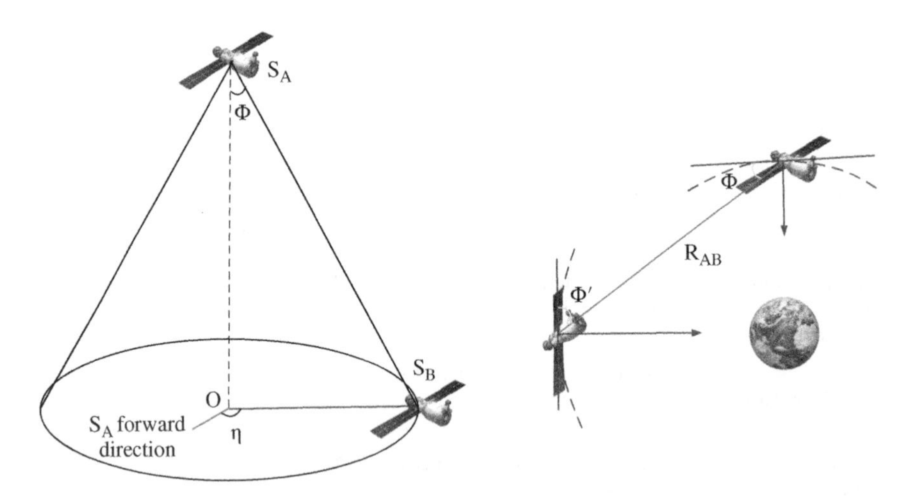

Figure 4.6 Schematic diagram of inter-satellite topology attributes.

Earth's geology blocks radio transmission, so satellites are considered invisible when the wire between them passes over the Earth. However, although the propagation loss of high-frequency microwaves in free space is small, severe loss occurs when passing through the atmosphere. Therefore, when calculating the influence of the Earth's occlusion on the visibility of the inter-satellite link, the influence of the atmosphere needs to be taken into account.

Assuming that the orbital height of the source satellite S_1 is d, the orbital height of the target satellite S_2 is d_2, and the inter-satellite distance is d, then the angle β between the tangent line when the link is tangent to the Earth and the normal direction of the source satellite antenna is

$$\beta = \arcsin\left(\frac{d_1}{d_E}\right) \tag{4.2}$$

Among them, d_E is the radius of the Earth. Since the addition of the atmospheric model is considered, Eq. (4.2) should be changed to

$$\beta = \arcsin\left(\frac{d_1}{d_E - d_g}\right) \tag{4.3}$$

where d_g is the thickness of the atmosphere.

The angle between the link and the normal direction of the source satellite antenna is

$$\theta = \arccos\left(\frac{(d_1^2 + d^2 - d_2^2)}{2dd_1}\right) \tag{4.4}$$

If the target satellite S_2 is located within the visible range of the source satellite S_1, the angle θ between the link and the antenna of the source satellite should satisfy the following relationship:

$$\beta < \theta < \alpha \tag{4.5}$$

Among them, α is the scanning range of the narrow beam antenna.

The visibility between satellites must not only satisfy that the target satellite is visible to the source satellite, but also must satisfy that the source satellite is also within the visible range of the target satellite. Therefore, only the above relationship is not a sufficient and necessary condition for the visibility between satellites S_1 and S_2, in addition, it must also meet:

$$\beta' < \theta' < \alpha \tag{4.6}$$

Among them, θ' is the angle between the inter-satellite link and the target satellite antenna, and β' is the angle between the inter-satellite link and the Earth's surface (including the atmosphere) when it is tangent to the target satellite antenna.

Therefore, only if both Eqs. (4.5 and 4.6) are satisfied, it can be said that the two satellites are mutually visible.

As shown in Figure 4.7, when there is Earth occlusion between S_1 and S_2, the two sides are invisible (Figure 4.7a); when S_1 and S_2 are limited by the scanning range of the antenna, they are invisible (Figure 4.7b); when only one of S_1 and S_2 exists in the visible range of the other party, and the other party does not exist in the visible range of the other party, they are still invisible (Figure 4.7c). Only when the two satellites are within the range, it can be said that the two satellites are visible, and such a link has the possibility of communication and measurement.

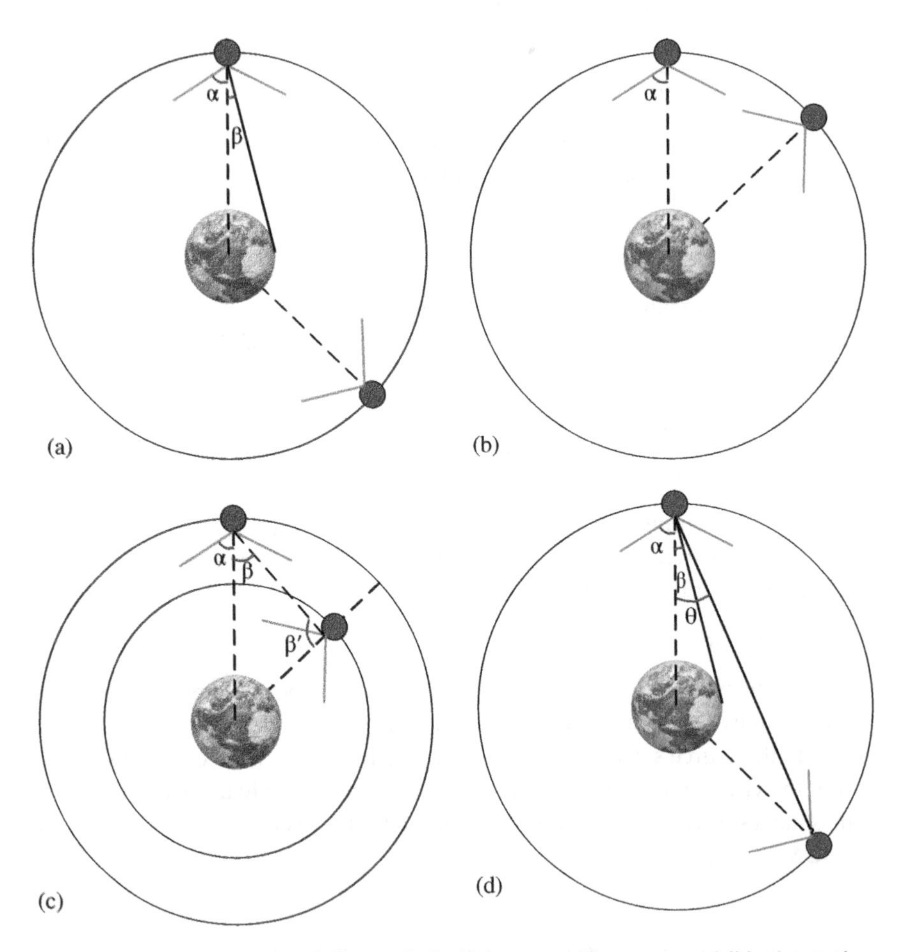

Figure 4.7 Inter-satellite visibility analysis. (a) Inter-satellites are not visible due to the Earth's occlusion; (b) invisibility between satellites due to antenna constraints; (c) only one satellite is invisible due to antenna constraints; (d) satellites can build links normally.

4.3.3 Inter-Satellite Link Topology Cost

In a communication network, topology refers to the schematic diagram of the geometric shape of the network, and the topology model that forms the network is the physical basis for building the network. The physical topology of a network is the geometric layout of the actual node-to-node connections. In a satellite constellation inter-satellite link network, nodes are all satellites that make up the constellation, and the connections between nodes are inter-satellite links. In traditional communication networks, common network topologies include bus, ring, star, mesh, etc. The structure diagram is shown in Figure 4.8. Due to the real-time characteristics of terrestrial networks, these topologies only contain connections in node space [11, 14].

This paper studies the network system based on STDMA; in addition to the similarities of spatial isolation with these topologies, the nodes connected to different targets at different times are also different. Taking the satellites on a certain orbital plane as an example, assuming that the nodes are visible to each other, Figure 4.9 describes a network topology of the combination of ring and star for these satellites in the TDMA access mode. The lines marked by A, B, and C in the figure represent different link connections between satellites in three different time slots, respectively.

In the STDMA inter-satellite link network, the connection objects and order of nodes directly determine the tasks and performance of the inter-satellite link. Therefore, different topology models will inevitably affect the transmission quality of the inter-satellite link. The impact on inter-satellite performance is mainly reflected in two aspects: communication quality and measurement accuracy. For a satellite network with n nodes, the loss caused by the topology model is

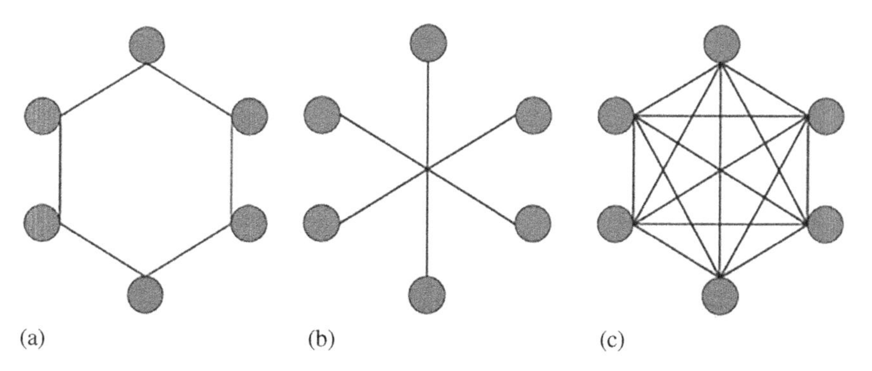

Figure 4.8 Common network topology diagram. (a) Ring topology; (b) star topology; (c) mesh topology.

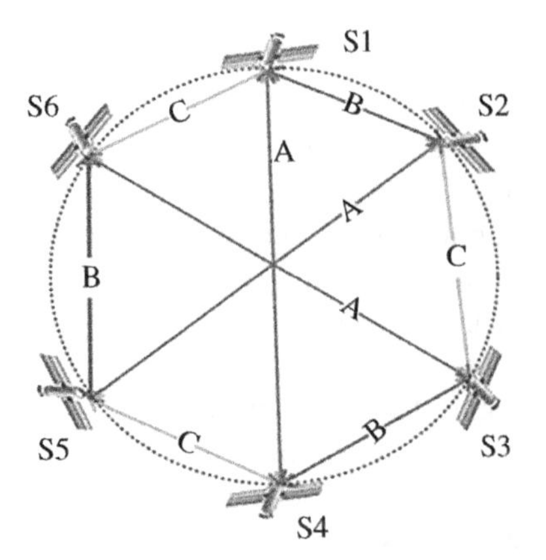

Figure 4.9 Schematic diagram of partial mesh topology in STDMA mode.

the average of the communication loss between the ground station and all satellites, and then the cost equation is:

$$\begin{cases} \overline{C_{trans}} = \sum_{j=1}^{n,\, j\neq i} C_{trans}(i,\ j)/(n-1) \\ \overline{C_{msr}} = \sum_{j=1}^{n,\, j\neq i} C_{msr}(i,\ j)/(n-1) \end{cases} \tag{4.7}$$

Among them, C_{trans} represents the communication cost from satellite i to satellite j, C_{msr} represents the measurement cost from satellite i to satellite j, and n represents the number of satellite nodes in the entire network.

This paper analyzes the advantages and disadvantages of the inter-satellite link network topology model from the perspective of the former, that is, the communication quality. The communication loss of a satellite link between satellites can be divided into three main parts: path loss, transmission loss, and protocol overhead. For a single satellite, the loss equation is:

$$C_{trans}(i, j) = \left(C_{path} + C_{antenma}\right) \times C_{protocal} \tag{4.8}$$

In this paper, the three parts of communication loss are converted to the angle of communication delay for analysis.

4.3.3.1 Path Loss

Path loss refers to the time delay caused by data propagation during communication and the time delay waiting on the relay satellite. Its equation can be written as:

$$C_{path} = t_{path} + t_{wait} \tag{4.9}$$

Among them, t_{path} is the delay caused by the link length, and t_{wait} is the time that the data waits on the relay satellite. Assuming that the total distance that the data travels in the air is d, then the propagation delay of the data in the vacuum is:

$$t_{trans} = d/c \tag{4.10}$$

where c is the speed of light in a vacuum, which is 3×10^8 m/s.

4.3.3.2 Transmission Loss

The transmission loss mainly refers to the data error probability caused by the transmission power of the antenna and the carrier-to-noise ratio of the signal reaching the destination. When the data error rate is high, the communication delay will inevitably increase. The link budget of the inter-satellite link can well reflect the loss of the space signal in transmission. The carrier-to-noise ratio of a signal reflects the quality of the signal propagating in space, and its expression is:

$$C/N_0 = EIRP - L_f - L_r - k + G_r/T_s \tag{4.11}$$

where $EIRP$ is the omnidirectional radiated power of the transmitting antenna, L_f is the free space propagation loss, L_r is the sum of the antenna angle loss and feeder loss, k is the Boltzmann constant, T_s is the receiver noise temperature, and G_r/T_s is the receiver quality factor.

Microwaves propagate in free space and their transmission loss mainly depends on the distance of propagation. The farther the distance, the greater the loss, and vice versa. The free space transmission loss of microwave is expressed as:

$$L_f = 10 \log \tag{4.12}$$

Among them, d is the signal propagation distance, f is the carrier frequency, and c is the speed of light.

In addition to the loss of microwave transmission in free space, the pitch angle of the inter-satellite link determines the angle between the direction of microwave transmission and reception and the direction of the maximum power of the antenna. Therefore, the size of the pitch angle will also play a certain role in the loss of microwave transmission. Assuming that the angle loss is linearly related to the angle, and the loss obtained by the satellite antenna due to the pitch angle is 0.05 dB/deg, and the feeder loss of the antenna is always 2 dB, since both the transmitting antenna and the receiving antenna have losses, then L_r can be written as

$$\begin{aligned} L_f &= L_e + L_k \\ &= 0.05(e_1 + e_2) + 4 \end{aligned} \tag{4.13}$$

Among them, L_e is the angle loss of the antenna, L_k is the feeder loss of the antenna, and e_1 and e_2 are the relative pitch angles of the transmitting satellite and the receiving satellite, respectively.

The relationship between *EIRP* and the power of the transmitting antenna is:

$$EIRP = P_t + G_t \tag{4.14}$$

Among them, P_t is the single-channel power of the transmit antenna, and G_t is the gain of the transmit antenna. Since the inter-satellite link uses a narrow beam antenna, the power of the transmitting antenna can be considered to be higher than that of the ordinary wide beam antenna. If we set the transmit power of the antenna to be 40 dBm and the antenna gain to be 26 dB, then the *EIRP* value can reach 37 dBW.

Assuming that the equivalent noise temperature of the receiver is 800 K, and the gain of the receiving antenna is the same as that of the transmitting antenna, then the receiver quality factor G_r/T_s is:

$$\begin{aligned}\frac{G_r}{T_s} &= 26\ \text{dB} - 10\ \log\ 800\ \text{dBK} \\ &= -7.03\ \text{dB/K}\end{aligned} \tag{4.15}$$

The link budget range of the final inter-satellite link is shown in Table 4.3.

Assuming that the communication rate of the inter-satellite link is R_b(bps) during data transmission, then its signal-to-noise ratio E_b/N_0 is:

$$\frac{E_b}{N_0} = \frac{C}{N_0} - 10\ \log R_b \tag{4.16}$$

Table 4.3 Inter-satellite link budget table.

	Value
Omnidirectional transmit power EIRP (dBW)	37
Inter-satellite link distance d (km)	7000–68,000
Free space propagation loss L_f (dB)	196–216
Antenna angle and feeder loss L_r (dB)	4–10
Receive antenna gain G_r (dB)	26
Equivalent noise temperature T_s (K)	800
Receiver quality factor G_s/T_s (dB/K)	−7.03
Receiver carrier-to-noise ratio C/N_0 (dB)	42.5–54.3

In this paper, it is assumed that LDPC coding is used, and the coding gain is set to $G_{code} = 9$ dB, then the data error rate of the inter-satellite link under BPSK modulation is:

$$P_{BER} = Q\left(2\sqrt{\left(\frac{G_{code} + E_b}{N_0}\right)}\right) \tag{4.17}$$

Among them, $Q(x)$ is the error complement function, and P_{BER} is the bit error probability. As can be seen from Table 4.3, taking the carrier-to-noise ratio as the minimum value of 42.5 dB/Hz, the relationship between the communication rate and the bit error rate is shown in Figure 4.10. It can be found from the figure that when the communication rate is close to 100 kbps, the bit error rate has been significantly increased, reaching the order of magnitude of 10^{-4}.

For a frame of 1024 bits data, the data frame error probability is:

$$P_{error} = 1 - (1 - P_{BER})^{1024} \tag{4.18}$$

The error of the data will bring about the retransmission of the data. Each retransmission requires the timeout of the retransmission timer. Assuming that the timeout time of the timer is t_c, the transmission loss can be expressed as the average delay caused by the data error, whose expression is:

$$C_{antenna} = t_{error} = P_{error} \cdot t_c \tag{4.19}$$

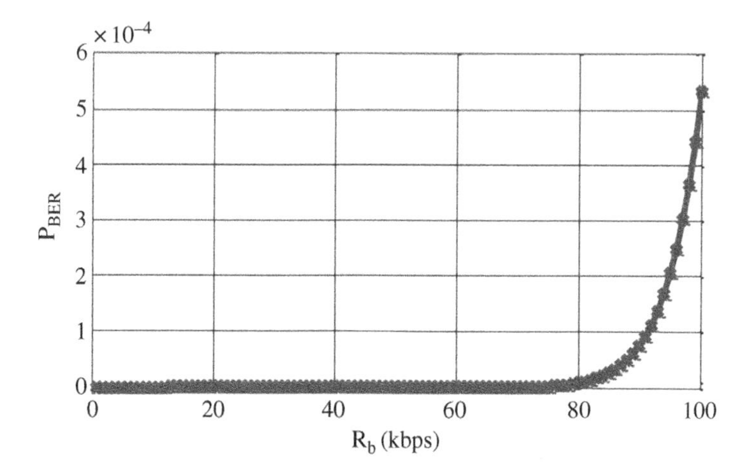

Figure 4.10 The relationship between the communication rate R_b(kbps) and the bit error rate P_{BER} at a carrier-to-noise ratio of 42.5 dB/Hz.

4.3.3.3 Protocol Overhead

The protocol overhead reflects the total length of the valid data in each frame of data. The expression of the protocol overhead is:

$$C_{protocal} = \frac{1}{1-\eta_p} \tag{4.20}$$

Among them, η_p is the ratio of the communication protocol length to the total frame length.

To sum up, the communication loss of the inter-satellite link consists of three parts: path loss, transmission loss, and protocol overhead. The following is a calculation and analysis of the inter-satellite link network of the topology model as shown in Figure 4.9. It is assumed that there are only one satellite S_1 and S_2 in the domestic satellite at this time, and the rest of the satellites need to be relayed through S_1 and S_2. Inter-satellite link timeslot division is performed in the order of A-B-C-A in the figure, the time slot is 1.5 seconds long, the link rate is 100 kbps, the retransmission timing t_c is 60 seconds, and η_p is 3%. Table 4.4 shows the calculation results of the communication loss under this topology model. Here, the cost of the ground station injected to the satellite is analyzed.

As can be seen from the table, since S_1 and S_2 are domestic satellites, there is no overhead in terms of inter-satellite links on the ground; satellites S_4 and S_5 are forwarded from time slot A by satellites S_1 and S_2, and the waiting delay is 1 time slots, the propagation delay is about 170 ms; satellites S_3 and S_6 are forwarded from time slot C by S_1 and S_2, respectively, the waiting delay is 1 time slot, and the propagation delay is about 160 ms; due to the link budget of these links are relatively large, so the transmission loss is relatively small here, all below 10 ms, so the final communication delay is mainly the delay caused by path loss, with an average of about 1125 ms.

Table 4.4 Communication loss budget of some mesh topology stars.

Satellite number	C_{path}	$C_{antenna}$	C_{trans}
S_1	0	0	0
S_2	0	0	0
S_3	1586.7 ms	7.5 ms	1644.4 ms
S_4	1673.3 ms	5.9 ms	1731.1 ms
S_5	1673.3 ms	5.9 ms	1731.1 ms
S_6	1586.7 ms	7.5 ms	1644.4 ms
Average	1086.7 ms	4.5 ms	1124.9 ms

4.4 Inter-Satellite Link Network Protocol Model

4.4.1 Inter-Satellite Network Protocol Model

The OSI model is a theoretical model of computer networks established by the International Organization for Standardization (ISO). In general, the OSI model is divided into seven layers, from the upper layer to the lower layer: application layer, presentation layer, session layer, transport layer, network layer, data link layer, and physical layer. Currently, the OSI model is used to design its own protocols for different layers according to the current network characteristics. In the terrestrial network, the TCP/IP protocol is the most used network protocol, which is mainly aimed at the network layer and the transport layer. Its simple and practical advantages make it widely used in the Internet [13, 15].

Because the number of nodes in the space network is greatly reduced compared with the ground network, and the transmission environment is also quite different from the ground network, in order to simplify the network model. The chapter refers to a spatial network model proposed by CCSDS to simplify and improve the OSI network model, making it more consistent with the requirements of the inter satellite link network designed in this book.

The function of the presentation layer in the ground network is to complete the "translation" problem caused by different coding and decoding in different types of hosts. At the same time, it can also encrypt and decrypt the transmitted data through the presentation layer. Since the spatial network in this paper assumes all nodes use the same encoding and do not involve cryptographic operations, so the presentation layer can be considered unnecessary in spatial networks. At the same time, the role of the session layer is to determine the initiation and interruption of communication. Since in the STDMA transmission mode of this paper, the initiation and interruption of communication are determined by the generation and end of the TDMA time slot, they have nothing to do with the transmission content and protocol. Therefore, the session layer is also not necessary in the spatial network in this paper.

Therefore, the protocol model for the inter-satellite link network designed in this paper is divided into five layers, as shown in Figure 4.11, from the upper layer to the lower layer are the application layer, transport layer, network layer, data link layer, and physical layer. The functions of these five layers are basically similar to the OSI model of the terrestrial network. The application layer refers to the specific application and operation of the transmitted content. The transport layer is responsible for numbering, classifying, controlling the data flow and ensuring the reliability, order, and error-free data of the transmitted content. During transmission, the transport layer is also the most important part of the network protocol. It can be understood as the control of the "order" of the transmitted data.

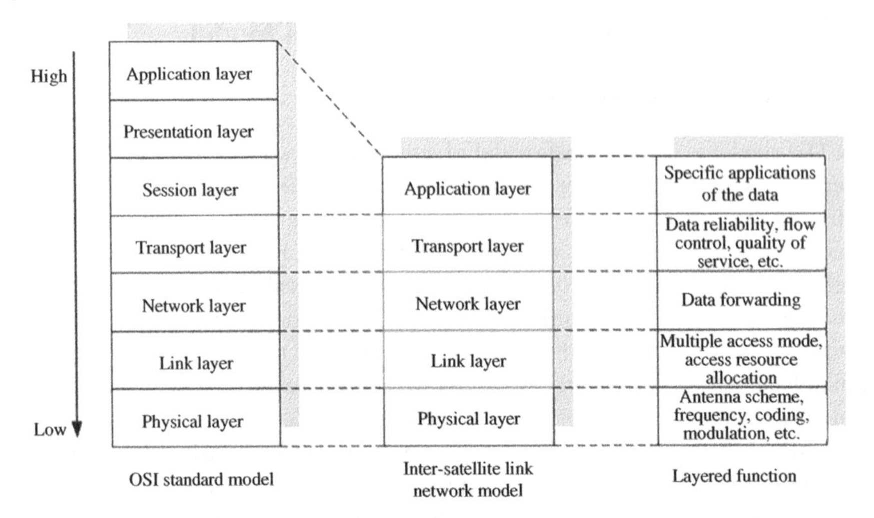

Figure 4.11 Inter-satellite link network layer model.

The network layer is specifically reflected in the addressing of each node in the network and the selection of data transmission paths. The data link layer in the STDMA network mode can be equivalent to the allocation of TDMA time slots based on the existing topology, the medium and interface of the physical layer value to transmit information, and the selection of antennas and working frequency bands in satellite networks [13].

The main research content of this paper mainly focuses on the link layer, network layer, and transport layer of the network. The link layer is the time slot allocation problem under STDMA network; the network layer is the data-forwarding strategy. The network layer is the data forwarding strategy; The transport layer is a transport protocol to ensure data reliability.

4.4.2 Transport Layer Protocol

Among the five-layer inter-satellite network protocols, the physical layer and application layer protocols are not within the scope of this paper, and their functions have also been described in the previous section, respectively. The data link and network layer will be designed in detail in Chapter 6, and the rest of the transport layer, as the core part of the protocol stack, has powerful functions and complex structures.

The functions of the transport layer protocol mainly include:

1) Data reliability assurance: including data response, data retransmission, retransmission timer setting, etc.

2) Network congestion processing and traffic adjustment mechanism: it is mainly used to control and adjust data transmission when the network is congested, and maintain network transmission performance.

In the terrestrial network, the network transport layer protocol generally uses the TCP protocol, and the application of the TCP protocol in the Internet has achieved great success. However, there are huge differences in nature between space transmission networks and terrestrial networks. The TCP protocol cannot be directly applied to the space communication network. The US Space Technology Advisory Committee CCSDS has made a series of modifications to the TCP protocol according to the characteristics of the space network, which is called the SCPS-TP protocol.

References

1 Chan, S. (2005). Architecture for a space-based information network with shared on-orbit processing. Massachusetts Institute of Technology, 154–170.

2 Quintas, D. and Friderikos, V. (2012). All optimal solutions in STDMA scheduling. *20th European Signal Processing Conference*. Bucharest, 834–838.

3 http://www.ccsds.com.

4 Ollie, L., Larry, B., Art, G. et al. (2005). GPS III system operations concepts. *IEEE Aerospace and Electronic Systems Magazine* 20 (1): 10–18.

5 CCSDS 714.0-B-2. Space communication protocol specification (SCPS) – transport protocol.

6 CCSDS 133.0-B-1. Space packet protocol.

7 Space Communication Architecture Working Group (2006). NASA space communication and navigation architecture recommendations for 2005-2030. *NASA*, 18–22.

8 Chang, S., Kim, B.W., Lee, C.G. et al. (1998). FSA-based link assignment and routing in low-earth orbit satellite networks. *IEEE Transactions on Vehicular Technology* 47 (3): 1037–1048P.

9 Lee, J. and Kang, S. (2000). Satellite over satellite (SOS) network: a novel architecture for satellite network. *Proceedings of IEEE INFOCOM 2000*, 1, 315–321.

10 Knoblock, E.J., Konangi, V.K., and Wallett, T.M. (2000). Comparison of SAFE and FTP for the South Pole TDRS relay system. *18thAIAA International Communications Satellite Systems Conference*, Oakland, CA (13 April 2000).

11 Vergados, D.J., Sgora, A., Vergados, D.D. et al. (2010). Fair TDMA scheduling in wireless multihop networks. *Telecommunication Systems* 50: 181–198.

12 Bhasin, K.B., Hackenberg, A.W., Slywczak, R.A. (2006). Lunar relay satellite network for space exploration: architecture, technologies and challenges. *24th*

AIAA International Communications Satellite Systems Conference (ICSSC), San Diego, CA.

13 Wallett, T.M. (2009). A brief survey of media access control, data link layer, and protocol technologies for lunar surface communications. *NASA, Glenn Research Center*, Cleveland, OH.

14 Li, X.B., Wang, Y.K., and Chen, J.Y. (2013). Time delay compensation of IIR notch filters for CW interference suppression in GNSS. *Proceedings of 2013 IEEE 11th International Conference on Electronics Measurement and Instruments*, 106–109.

15 Duan, Z.L. and Liu, H.Y. (2009). Analysis on navigation receivers error sources. *Radio Engineering* 39 (7): 37–40.

5

Principles of Laser Inter-Satellite Ranging

5.1 Principle of Inter-Satellite Ranging

The inter-satellite distance measurement and clock error measurement use the bidirectional incoherent information frame spread spectrum measurement system. The specific principle and timing relationship are shown in Figure 5.1.

In the figure, satellite A and satellite B are, respectively, based on their own clocks, and transmit ranging signals in their respective allocated time slots. Due to the inconsistency of the time between the two parties, there is a time difference between the ranging information frames sent by the two parties. In satellite A, the transmission delay can be obtained by capturing and tracking the ranging signal transmitted by satellite B. In addition to the electromagnetic wave propagation delay, it also includes the delay of the transmitting equipment of satellite B, the delay of the receiving equipment of satellite A, the clock difference between satellite B and satellite A, and the measurement error. The relationship is:

$$\rho_A = \Delta t + t_B + \tau_{BA} + r_A + \varepsilon_A \tag{5.1}$$

In the same way, satellite B can measure the time delay ρ_B, and its time relationship:

$$\rho_B = (-\Delta t) + t_A + \tau_{AB} + r_B + \varepsilon_B \tag{5.2}$$

If the ranging frame sent by satellite B is regarded as a radio ruler between satellite A and satellite B, the epoch of the transmission frame is the radio ruler scale, which can measure the geometric distance between the two satellites at a certain moment. The difference from measuring the distance between two objects under conventional static conditions is that there is mutual movement between the two satellites and between the satellite and the radio ruler. At this time, an ideal "sampling shutter" is needed to simultaneously extract the scale of the radio wave size at a certain moment in satellite A and satellite B, and calculate the geometric distance of the two satellites in an inertial reference system/UTC time system [1].

Laser Inter-Satellite Links Technology, First Edition. Jianjun Zhang and Jing Li.

Published 2023 by John Wiley & Sons, Inc.

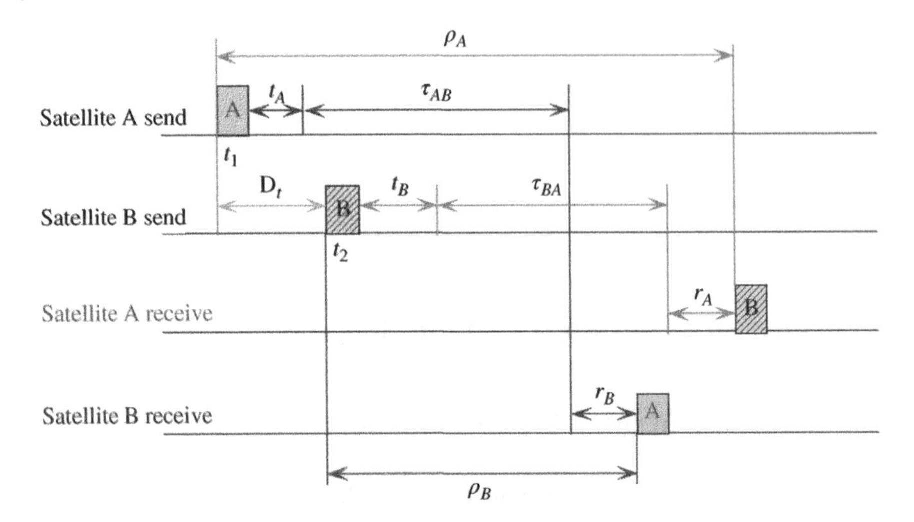

Figure 5.1 The principle and timing relationship of two-way incoherent information frame ranging and clock difference measurement.

5.2 Inter-Satellite Ranging Accuracy

The inter-satellite link space ranging performance is related to the specific ranging terminal type. For example, the phase center and channel delay of a reflector narrow-beam antenna suitable for establishing a static inter-satellite network are related to the antenna structure, and the position of the phase center in the satellite coordinate system changes with the rotation of the antenna. A phased array antenna suitable for establishing a dynamic inter-satellite network has no moving parts, and its phase center is considered to be fixed, but the channel delay characteristics will vary with the beam pointing. In addition, the delay characteristics of other devices in the inter-satellite link channel and the influence of different temperature conditions on the ranging accuracy are also different [2, 3].

The object of the inter-satellite link measurement is the real distance (i.e. true distance) between the satellite centroids. The direct measurement value is the pseudo-range calculated by the inter-satellite transceiver, and it is necessary to deduct the zero value of the channel delay and correct the change of the phase center, correct the clock error and other factors to get the pseudo-range:

$$\tau_{ij} = \tau_i^{clk} + \tau_i^0 - \underbrace{\tau_i^{pcv} + \tau_{ij}^{true} - \tau_j^{pcv}}_{\tau_{ij}^{pc}} + \tau_j^0 - \tau_j^{clk} \tag{5.3}$$

Among them, the subscripts i and j represent the satellite number, the superscript true represents the true distance between the two sides, the superscript *pc* represents the phase center distance between the two sides, the superscript 0 represents the channel delay zero value, the superscript *clk* represents the satellite clock difference, and the superscript *pcv* represents phase center variation.

It can be seen from the above formula that the ranging accuracy is the accuracy of the pseudo-range τ_{ij}, and the factors affecting the ranging accuracy are mainly the accuracy of the delay zero value representing the delay characteristics of the ranging channel, and the accuracy of the phase center variation representing the phase center characteristics of the antenna. The ranging channel refers to the ranging signal transmission channel including transceivers and antennas. For one end of the link, the delay zero value of the ranging channel starts from the baseband processing of the ranging signal and ends at the antenna phase center. The phase center of the antenna is related to the structure of the satellite, and the phase center variation (*pcv*) refers to the projection length of the line connecting the phase center and the satellite centroid in the ranging direction. In practical applications, it is assumed that the time delay zero value is a constant, and the phase center change is a variable, which is predicted by the onboard software.

5.3 Principle of Microwave Inter-Satellite Ranging

5.3.1 Principle of Pseudo-Range Two-Way Ranging

Microwave signals travel in a vacuum at the speed of light. When a satellite transmits a microwave signal to another satellite, if we know when the two satellites are sending and receiving the signal, then we can get a measurement between the two satellites' distance, which is called pseudo-range, as shown in Figure 5.2.

Assuming that the signal transmission time of a transmitting satellite is t_s, the signal receiving time of the receiving satellite is t_R, c is the speed of light. Without considering the satellite clock difference, the pseudo-range measurement between the two satellites is

$$\rho(X_s, X_R) = c \times (t_R - t_s) \tag{5.4}$$

However, there is a clock difference between the satellite-borne clocks, and the measurement accuracy of the pseudo-range is greatly affected by the clock difference. At present, the best way to deal with this error is the satellite two-way measurement method. Inter-satellite link bidirectional ranging is a process in which two satellites alternately act as transmitting satellites and receiving satellite transmission signals. It is composed of two signal transmissions. The first satellite i acts as transmitting satellites and satellite j acts as receiving satellites. The second signal transmission, satellite i acts as receiving satellites and satellite j acts as transmitting satellites. The two-way ranging process is shown in Figure 5.3.

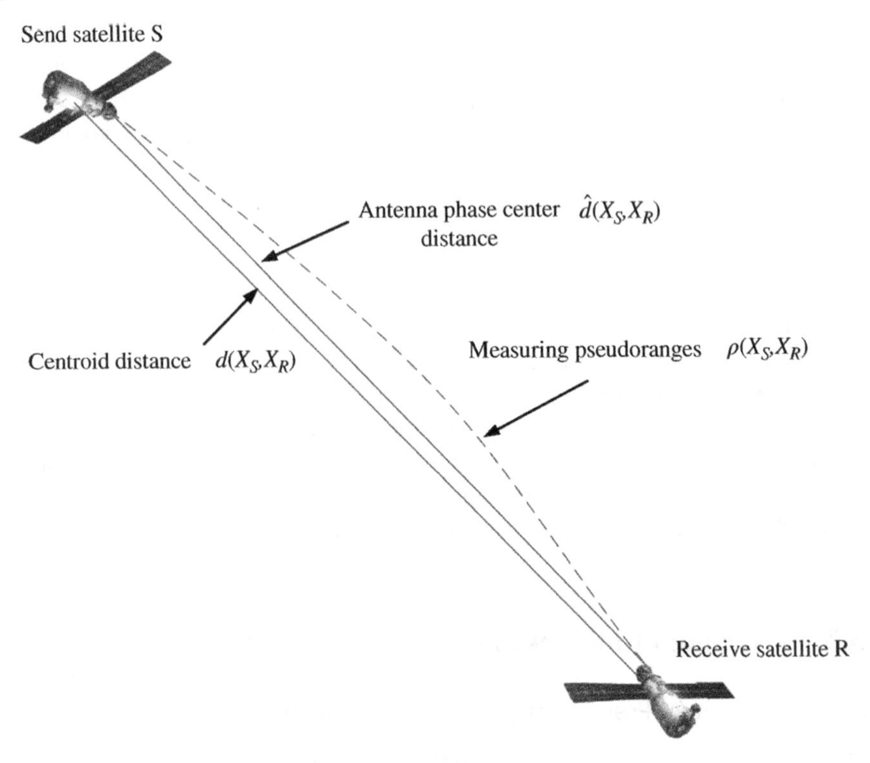

Figure 5.2 Schematic diagram of pseudo-range measurement.

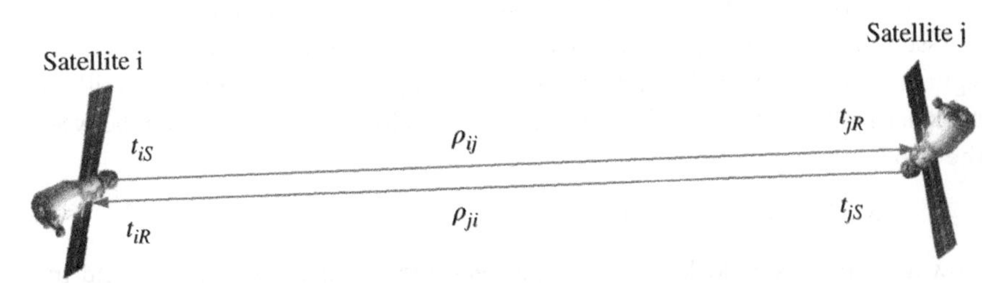

Figure 5.3 Schematic diagram of two-way measurement between satellites.

Satellite i sends a microwave signal to the satellite j when it is t_{is} on the onboard clock face; after the space delay of signal transmission τ_{AB}, it is received by the satellite j at the time when the satellite is t_{jR} on the satellite onboard clock face. On the other hand, the satellite j also sends a ranging signal to the satellite at the time t_{jS} of the clock face, and is received by the satellite i at the time t_{iR} of the satellite's clock face. Through two one-way measurements, the pseudo-ranges measured twice can be obtained according to formula (5.4), as follows:

$$\rho_{ij} = c(t_{jR} - t_{is}) = R_{ij} + c\Delta t_{ij} - \Delta L_{ij} \tag{5.5}$$

$$\rho_{ji} = c(t_{iR} - t_{js}) = R_{ji} + c\Delta t_{ji} - \Delta L_{ji} \tag{5.6}$$

Among them, ρ is the pseudo-range of the space measurement, R is the spatial geometric distance between the two satellites, Δt_{ij} and Δt_{ji} are the clock difference existing between the two satellites, which are opposite numbers to each other, while ΔL is the measured distance error of the two satellites during the measurement process, such as antenna phase center error, multipath effect error, relativistic effect error, etc. [4].

Add formulas (5.5) and (5.6) and divide by 2 to get

$$\frac{\rho_{ij} + \rho_{ji}}{2} = R - \frac{\Delta L_{ij} + \Delta L_{ji}}{2} \tag{5.7}$$

which is $R = R_{ij} = R_{ji} = \sqrt{(x_i - x)^2 + (y_i - y)^2 + (z_i - z)^2}$.

It can be seen from formula (5.7) that the two-way ranging method can offset the error caused by the clock difference between satellites, but the ranging error ΔL needs to establish and verify the error model according to various error sources.

5.3.2 Analysis of Error Sources in Microwave Ranging

The main source of errors in microwave ranging:

1) Errors related to satellite structure, including antenna phase center error, equipment circuit delay error, equipment noise error, etc.
2) Errors in the signal transmission process, mainly including multipath effect errors, ionospheric delay errors, etc.
3) Other errors, mainly relativistic effect errors, etc.

Among the above errors, some errors usually have certain rules to follow, which can be corrected by mathematical models, while noise errors have large randomness and can only be compensated by statistical methods [5].

5.3.2.1 Antenna Phase Center Error

In the inter-satellite link measurement process, the pseudo-range is the distance between the corresponding satellite antenna phase centers, but in the satellite orbit determination process, the orbit dynamics model used in orbit extrapolation and the state information in orbit estimation are both based on the satellite's centroid, which results in a positional error between the satellite's centroid and the antenna's phase center. As shown in Figure 5.4, after analyzing the geometric

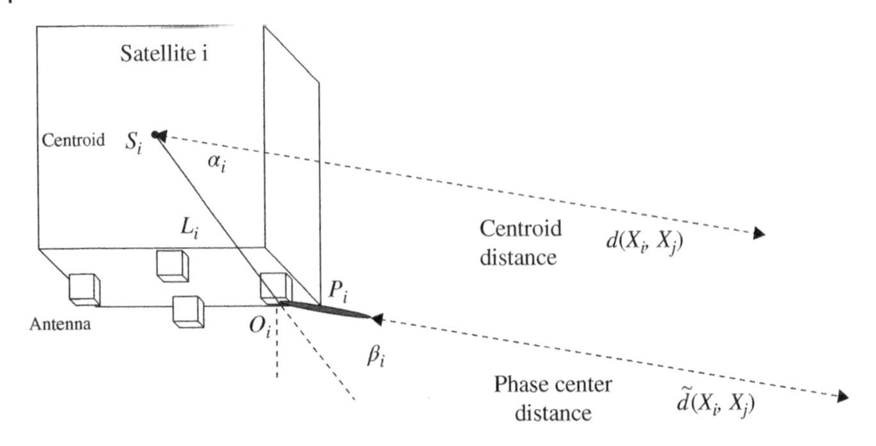

Figure 5.4 The relationship between satellite centroid and phase center in measurement.

relationship between the antenna phase center and the satellite centroid, the distance between the two satellite antenna phase centers can be converted to the distance between the two satellites' centroids [6].

In the figure, L_i is the distance from the satellite i's center of mass to the antenna phase zero O_i, angle α_i is the angle between and the line connecting the two satellites' centroids, angle β_i is the antenna beam azimuth, P_i is the antenna phase zero to the antenna phase center. The distance between two satellites is expressed by the centroid distance and the phase center distance is:

$$d(X_i, X_j) = \tilde{d}(X_i, X_j) + \Delta\rho_{oF} \tag{5.8}$$

Correspondingly, the measurement of pseudo-range also needs to correct the antenna phase error.

$$\rho_{ij} = \tilde{\rho}_{ij} + \Delta\rho_{oF} \tag{5.9}$$

$\Delta\rho_{oF}$ is correcting distance, the calculation formula is:

$$\Delta\rho_{oF} = L_i \cos\alpha_i + P_i \cos(\alpha_i - \beta_i) + L_j \cos\alpha_j + P_j \cos(\alpha_i - \beta_i) \tag{5.10}$$

where satellite j represents another satellite connected to satellite i.

In the error calculation process, the distance L between the satellite mass center and the antenna phase zero point can be obtained during the satellite manufacturing process, angle α and β can be calculated according to the satellite attitude measurement. At present, the error of the satellite attitude measurement is less than 0.1 degrees. Therefore, the error will be less than 1 degree. Assuming that L is 1 m, the error of the correction distance $\Delta\rho_{oF}$ will be less than 8 mm.

5.3.2.2 Device Circuit Delay Error

The delay error of the signal transmitter and receiver equipment circuit mainly has two aspects: the clock jitter error when the baseband signal is output and the error caused by the change of the equipment environment (mainly the equipment temperature). Every 1 degree increase of the measuring equipment will bring delay, and generally the temperature change of satellite equipment is controlled at plus or minus 2 degrees, so the delay error will be less than 0.2 ns; generally, this error will be considered in the satellite clock error during the satellite signal transmission process [7].

5.3.2.3 Multipath Effect Error

In the process of signal transmission, the signal is often reflected by the equipment near the antenna to generate multipath signals. At this time, the signal received by the antenna will be a composite signal, which leads to the deterioration of the ranging accuracy. A typical multipath signal model is as follows:

$$\begin{aligned} s(t) &= s_d(t) + \sum s_i(t) \\ &= Ap(t)\sin(2\pi ft) + \sum[\alpha_i Ap(t-\tau)\sin(2\pi ft + \varphi_i)] \end{aligned} \tag{5.11}$$

Among them, A is the amplitude of the signal source, $p(t)$ is the XOR of the data code and the pseudo-code when the signal is measured and transmitted, f is the carrier frequency, α_i is the attenuation coefficient of the reflected wave, τ is the propagation delay of the reflected wave, and φ_i is the total phase change of the reflected wave.

Different measurement signals will have different multipath effects. The UHF signal has a large multipath due to its large antenna beam width, and the maximum error can reach 1.414. Although there are relatively mature mathematical models to compensate for the multipath effect error, in actual measurement, it is difficult to completely cancel it, and more is to cancel it with statistical significance.

5.3.2.4 Ionospheric Delay Error

When establishing inter-satellite links between satellites, under normal circumstances, the visibility between satellites avoids the ionosphere 1000 km above the Earth, but some of the links between satellites will still be close to the Earth and are affected by the thin ionosphere. However, the chain building between some satellites will still be close to the Earth, which is affected by the thin ionosphere. The ionospheric delay model is as follows:

$$\Delta\rho_I = \int_S (n_g - 1)dl = \frac{40.28}{f^2}\int_S n_e dl = 40.28\frac{N_e}{f^2} \tag{5.12}$$

where $N_e = \int_S n_e dl$ is TEC, that is, the total amount of electrons contained in the spatial propagation path of the tubular channel per unit cross-sectional area; from formula (5.12), it can be seen that the higher the signal frequency, the smaller the influence of the ionosphere, so the residual laser signal measurement error after cancellation will be much smaller than the UHF microwave signal.

5.3.2.5 Relativistic Effect Error

Signal transmission between satellites is affected by relativistic effects and requires error correction. The frequency change caused by the special and general relativity of the spaceborne atomic frequency standard is as follows:

$$\begin{aligned}\Delta f &= \Delta f_1 + \Delta f_2 = -\frac{v^2}{2c^2}f + \frac{\mu}{c^2}\left(\frac{1}{R} - \frac{1}{r}\right)f \\ &= \frac{f}{c^2}\left(\frac{\mu}{R} - \frac{\mu}{r} - \frac{v^2}{2}\right) \\ &= \frac{\mu}{c^2}\left(\frac{1}{R} - \frac{2}{a(1-e\cos E)} - \frac{1}{2\alpha}\right)f\end{aligned} \tag{5.13}$$

Among them, f is the frequency of the atomic frequency scale at rest in the inertial reference frame, v is satellite speed, e is Earth's gravitational constant that generally takes $3.98600436 \times 10^{14} \text{m}^3/\text{s}^2$, a is satellite orbit semimajor axis, e is the orbital eccentricity, E is the orbital anomaly angle, R and r represent the directions of the signal receiver and signal source, respectively.

According to formula (5.13), the frequency difference caused by the relativistic effect of satellites whose orbits are approximately circular can be calculated:

$$\Delta f = 4.45 \times 10^{-10} f \tag{5.14}$$

Since the actual satellite orbit is generally an elliptical orbit, the clock error of the satellite needs to be compensated, as follows:

$$\Delta t_r = Fe\sqrt{a}\sin E \tag{5.15a}$$

where $F = -4.4428076321 \times 10$.

When mutual measurement is performed between the satellites i, j, the satellites are in relative motion. Due to the relativistic effect, it can be known from formula (5.14) that the frequency deviation caused by the satellite is:

$$\Delta f = \frac{f_i \mu}{c^2}\left(\frac{1}{r_j} - \frac{1}{r_i}\right) + \frac{f_i}{2c^2}\left(v_j^2 - v_i^2\right) \tag{5.15b}$$

The time offset is

$$\Delta t = \frac{t_i \mu}{c^2}\left(\frac{1}{r_j} - \frac{1}{r_i}\right) + \frac{t_i}{2c^2}\left(v_j^2 - v_i^2\right) \tag{5.16}$$

When the satellite orbit is the Walker constellation orbit that tends to be circular, the relative change between the two satellites is not obvious.

5.4 Principle of Laser Inter-Satellite Ranging

5.4.1 Principle of Laser Pulse Ranging

There are two main types of laser ranging methods, namely pulse measurement and phase measurement. Phase measurement is an ultra-high-precision measurement method with a short operating distance. In the navigation satellite system, the MEO orbital altitude is above 20,000 km, and the distance between the two satellites is high. The longest distance is more than 40,000 km. Therefore, the inter-satellite laser link should adopt the principle of pulse ranging [4, 6].

Pulse laser ranging is that the laser emits an optical pulse to the measured target, and the target captures the optical pulse and reflects the optical pulse back. After the receiver receives the optical pulse reflected from the target, the time difference is recorded and the distance between the two is calculated.

$$d = \frac{ct}{2} \tag{5.17}$$

The pulse laser ranging method has a very long range. The current Earth–Moon ranging system adopts the pulse ranging method, and the measurement distance can reach 400,000 km, which far exceeds the distance requirement for inter-satellite measurement. In addition, the accuracy of pulse ranging is related to the pulse width of the laser. According to ranging formula of laser error:

$$\Delta L = c\Delta t \tag{5.18}$$

It can be seen that the laser ranging error is theoretically proportional to the pulse width. The specific pulse width and measurement accuracy are shown in Table 5.1.

In 1964, when the laser ranging technology was just in its infancy, a laser with a pulse width of 10 ns was used. In the 1990s, the pulse width ranging of 100 ps has

Table 5.1 Comparison of laser ranging pulse width and ranging accuracy.

Pulse width	Measurement accuracy
10 ns	3 m
100 ps	3 cm
10 ps	3 mm

been used, and the current fourth-generation laser ranging technology can reach the ranging accuracy of 1 cm.

5.4.2 Analysis of Error Sources in Laser Ranging

Laser inter-satellite ranging mainly considers two error sources: relativistic error and satellite measurement center error. In the process of inter-satellite measurement, the laser measurement method is similar to the microwave measurement method. The signal propagates at the speed of light, and the signal source and receiver are also installed on the same satellite. Therefore, the error caused by the relativistic effect can be represented by the same model. The principle of satellite laser measurement center error is similar to that of microwave antenna phase center error. The position of the laser signal source or reflector is reduced to the satellite mass center [3, 7].

Assuming that the difference between the phase zero point and the phase center of the antenna of the laser signal source or reflector is not considered, and the signal source and the reflector are installed at the same position, the measurement center error of the laser can be expressed according to formula (5.12) as

$$\Delta\rho_{oFS} = Ls_i \cos\alpha_i + Ls_j \cos\alpha_j \tag{5.19}$$

Among them, Ls_i and Ls_j, respectively, represent the distances of the two satellite laser signal sources and reflectors from the satellite center of mass, α_i and α_j respectively, represent the angle between the line connecting the centroids of the two satellites and the line connecting the centroid and the signal source or reflector. Replacing the scalar formula (5.19) with the vector formula, expressing the position of the laser signal source or transmitter with the star-fixed coordinate system, and expressing the distance between the two satellites with the J2000 ground-fixed coordinate system, then we can get formula

$$\Delta\rho_{oFS} = \frac{[C_i](X_i - X_j)S_i}{|X_i - X_j|} + \frac{[C_j](X_i - X_j)S_j}{|X_i - X_j|} \tag{5.20}$$

Among them, $[C_i]$ and $[C_j]$, respectively, represent the transformation matrices from the ground-fixed coordinate system to the satellite-fixed coordinate system of the two satellites; X_i and X_j, respectively, represent the position state vectors of the two satellites in the J2000 ground-fixed coordinate system; S_i and S_j, respectively, represent the position vectors of the laser signal sources or reflectors of the two satellites in the satellite-fixed coordinate system to which they belong. Among the satellites currently installed with laser reflectors, the star-fixed coordinates of the GLONASS satellite are $S_{GLONASS} = [0.00\ \ 0.00\ \ 1.51]^T$, while the star-fixed coordinates of the GPS satellites are $S_{GPS} = [0.86\ \ -0.52\ \ 0.66]^T$.

References

1 Rani, M. and Prince, S. (2012). A study on inter-satellite optical wireless communication and its performance analysis. *2012 IEEE International Conference on, Devices, Circuits and Systems (ICDCS)*, 202–205. IEEE.

2 Petrovich, D., Gill, R.A., and Feldmann, R.J. (2000). Demonstration of a high-altitude laser crosslink. *Aerospace Conference Proceedings*, 67–77. IEEE.

3 Gregory, M., Heine, F., Kämpfner, H. et al. (2011). Coherent inter-satellite and satellite-ground laser links. *Free-Space Laser Communication Technologies XXIII* 7923 (6).

4 Feldmann, R.J. and Gill, R.A. (1998). Development of laser crosslink for airborne operations. *Military Communications Conference*, 633–637. IEEE.

5 Luba, O., Boyd, L., Gower, A. et al. (2005). GPS III system operations concepts. *IEEE Aerospaceand Electronic Systems Magazine* 20 (1): 10–18.

6 Maine, K., Anderson, P., and Bayuk, F. (2004). Communication architecture for GPSIII. *Proceedings of IEEE Aerospace Conference, Aerospace Corporation*, Los Angeles, CA, 124–129.

7 Maine, K.P., Anderson, P, and Langer, J. (2003). Crosslinks for the next-generation GPS. *IEEE Aerospace Conference Proceedings*, 4_1589–4_1596.

6

Composition of Laser Inter-Satellite Link

6.1 Basic Structure of Laser Inter-Satellite Link

The laser inter-satellite link system is composed of optical sending subsystem, optical receiving subsystem, alignment, acquisition, tracking subsystem, and some other auxiliary systems. The principle block diagram of the system composition is shown in Figure 6.1.

6.1.1 Optical Transmitting Subsystem

The optical transmitter subsystem is mainly composed of light source, modulator, and optical antenna.

The transmitting and receiving antenna of the laser inter-satellite link system is actually an optical telescope, and the type of the antenna can be a Cassegrain-type reflective antenna or a transmission antenna according to the specific situation.

At present, the more commonly used optical launch systems are small and medium-power semiconductor or solid-state lasers plus high-power low-noise fiber amplifiers to form the main vibration power amplifier system (MOPA).

In the LEO–LEO (low-orbit satellite–low-orbit satellite) and LEO–GEO (low-orbit satellite–geostationary orbit satellite) links in the United States, Europe, and Japan, the wavelengths of AlGaAs lasers in the range of 800–850 nm are used, because the APD detection devices work at the peak, with high quantum efficiency and high gain. In the ground device in the satellite-to-earth link, a frequency-doubling Nd:YAG laser or a hydrogen ion laser is used as the light source, with a wavelength of 514–532 nm. This band has strong anti-interference ability and can pass through the atmosphere without interrupting communication. With the development of semiconductor lasers, the light sources used in satellite optical communications in the future will tend to develop to shorter wavelength bands. The semiconductor laser-pumped Nd:YAG laser not only has good coherence, but also can be made small, so it is also a good choice for future onboard lasers [1].

Laser Inter-Satellite Links Technology, First Edition. Jianjun Zhang and Jing Li.

Published 2023 by John Wiley & Sons, Inc.

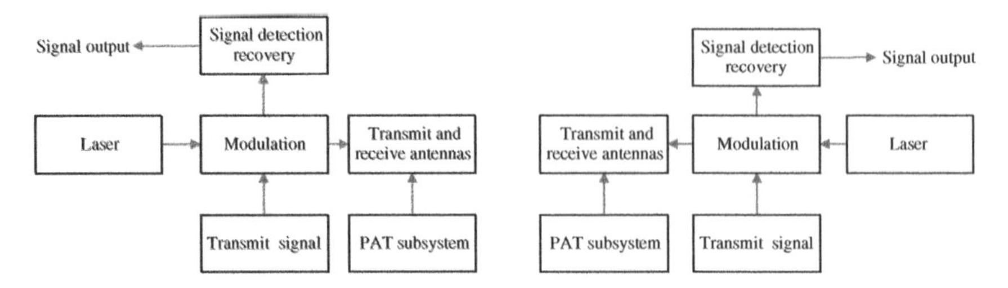

Figure 6.1 Block diagram of the laser inter-satellite link system.

Generally speaking, in the currently selected satellite optical communication band range, for antennas with larger apertures, such as the 25 cm antenna of the SILEX system, a reflective antenna can be used, which helps to reduce the difficulty of manufacturing the antenna and improve the reliability of the antenna. Reduce weight, and when the antenna aperture is small, a transmissive antenna is used, such as an antenna system for a small optical user terminal (SOUT).

Since the aperture of the antenna directly affects the gain of the antenna, the larger the aperture, the greater the gain. Therefore, from the perspective of improving the antenna gain, the antenna aperture of the satellite optical communication system should be larger. However, as the aperture increases, the volume and weight of the antenna will also increase, so the onboard antenna aperture should not be too large. Generally, the onboard antenna aperture of the satellite optical communication system is about 30 cm. For example, the SILEX system installed on the GEO satellite has an antenna aperture of 25 cm, and the one installed on the LEO is 18 cm. The receiving antenna aperture of the satellite optical communication system developed by JPL is 32 cm. The onboard antenna aperture of the satellite optical communication system of Japan's air-to-ground optical communication experiment is 30 cm.

6.1.2 Light Receiving Subsystem

The optical receiver subsystem consists of an optical receiving antenna, a detector, and a demodulator. In addition, the detection part also includes filtering, amplifying, and other parts.

Filters and detectors are important parts of the receiving system. At present, the filtering in the satellite optical communication systems developed by the United States, Europe, and Japan basically uses interference filters, and the half bandwidth is less than 7 nm, which helps to simplify the entire receiving system and improve the reliability of the system. At the same time, for the GEO–LEG link, due to the high relative motion speed between the two satellites, a large Doppler frequency shift will occur, so the bandwidth of the filter cannot be selected too narrow. In this regard, interference filters are also a good choice. Detectors that receive communication signals generally use avalanche photodiodes (APDs) because they have high gain and peak sensitivity around 800 nm.

Like other optical communication methods, the laser inter-satellite link can be a direct detection system and a coherent detection system. The coherent system is divided into two methods: heterodyne detection and homodyne detection. In the direct detection system, the modulated signal is obtained by directly modulating the intensity of the optical carrier at the transmitting end, and the optical signal is converted into an electrical signal by a photodetector at the receiving end. The simplest way of such a structure is to use on-off keying (OOK) intensity modulation, but pulse modulation (PPM) is more commonly used. Direct detection has the advantages of simple structure and low cost, but it has low-frequency band utilization and low receiving sensitivity [2].

6.1.3 Align, Capture, Track Subsystem (PAT)

Before the laser inter-satellite link system performs data transmission, the optical field power of the transmitter must first reach the detector of the receiver. The act of aiming the transmitter in the proper direction is called Pointing. The receiver operation to determine the direction of arrival of the incoming beam is called acquisition. The operation of maintaining alignment and acquisition throughout the communication period is called Tracking. Alignment, capture, and tracking subsystem is one of the very important subsystems in space laser communication system, and it is also the significant difference between space laser communication and other communication systems [3, 4].

6.2 Workflow of Laser Inter-Satellite Link

In the laser communication process, from the beginning to the end of the communication, the entire working process of the laser terminal can be divided into four stages: the initial pointing stage of the laser boresight, the scanning capture stage, the tracking stage, and the final communication establishment stage. The specific process of the laser system work is shown in Figure 6.2. In the whole workflow, each stage is forced to be interrupted for some reasons, such as the target is not within the visible range, the inter-satellite link changes, the task is cancelled, etc. After each task is completed, the interrupt machine enters a standby state until the next work instruction is received, and the above workflow needs to be restarted to execute a new task instruction.

The onboard platform issues a task command for communication, interrupting the machine to enter the initial pointing stage. The lasers, advance sights, servo

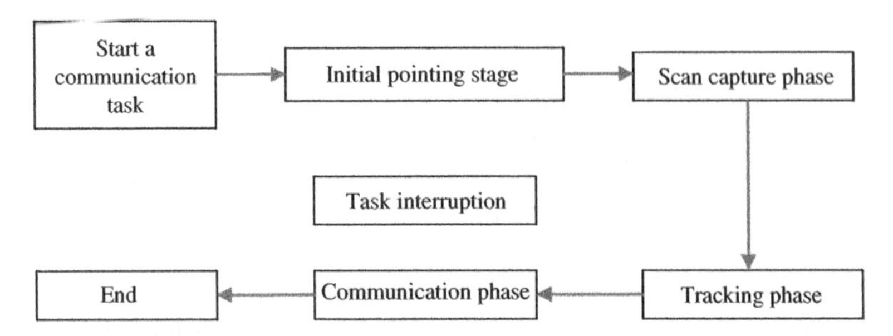

Figure 6.2 Workflow of laser inter-satellite link.

turntables, and other devices of the spaceborne laser terminal all perform self-checking work, enter the calibration mode, and the laser terminal enters the initial state. The control subsystem obtains ephemeris and other information, calculates the initial pointing of the target to be communicated, and controls the turntable to guide the laser boresight to the estimated position of the cooperative target in combination with the angle of the turntable itself [5, 6].

After the control turntable realizes the rotation of the boresight, the boresight is directed to the capture uncertainty area where the target to be linked is located, and the communication terminal enters the scanning capture stage at this time. According to the information data injection of the orbit forecast, the size of the captured uncertain area is calculated, the real-time scanning trajectory of the boresight is calculated, and the servomechanism is driven to scan the captured uncertain area. The transmitting signal of the satellite at the transmitting end is scanned by the laser, and the detector starts to detect the signal at the same time. When the response signal is detected, the direction of the boresight is adjusted again, so that the boresight enters the detection field of view of the receiving satellite terminal. At this time, it is ready to enter the signal tracking phase [7, 8].

The two satellite terminals that communicate with each other capture the initial signal, perform rough tracking, and continuously control and change the boresight of the laser. When the tracking accuracy meets the system requirements, the terminal can enter the fine tracking stage and establish a real-time and stable inter satellite link. When the tracking loop loses lock, it needs to return to the rough tracking in order or even to the scanning stage, and repeat the above process to re-establish the connection. After the step of precise tracking is achieved, the system can start to enter the communication phase after maintaining stable tracking. The terminal that transmits the satellite performs data encoding, modulation, and signal transmission, and the terminal that receives the satellite receives the signal light and performs photoelectric detection, demodulation, and decoding [9–11].

6.3 Constraints

The carriers of the two communication terminals for inter-satellite laser communication are satellites orbiting the Earth in different orbits. In order to successfully establish the communication link between the two mobile terminals and keep the communication link in a good state, the designed communication terminal must adapt to the dynamic environment in which it is located. Therefore, it is necessary to study the performance of satellite orbit and satellite platform, and obtain the dynamic range and environmental conditions that the laser communication terminal needs to adapt to when working. The relative motion relationship between satellites, the determination and prediction accuracy of the satellite orbit, the actual satellite load and the dynamic characteristics of the control system, and their mutual influence will determine the very important parameters in the terminal design, such as scanning strategy, subarea division, tracking servo bandwidth, installation layout, and dynamic characteristics [12].

6.3.1 Satellite Orbit

The various satellites that make up the inter-satellite link can be divided into three categories according to the orbit height: low-orbit satellite (LEO), medium orbit satellite (MEO), and geosynchronous orbit satellite (GEO). There are many types of links. Table 6.1 shows the basic parameters of the three main links. This topic mainly studies the link between low-orbit satellites and geostationary orbit satellites.

In the operation of the space optical communication link system, the accuracy of satellite orbit positioning and attitude determination is very important to ensure the normal operation of the system. Satellite optical communication terminals

Table 6.1 Basic link data.

Status parameter	GEO–GEO (θ = 120°)	LEO–GEP (h = 250 km)	LEO–GEO (h = 800 km)
Distance	$R = 73{,}160$ km	$35{,}620 < R < 43{,}560$ km	$35{,}070 < R < 45{,}050$ km
Angle coverage	$0.2° \times 0.2°$	$18.6°\ 2^{\pi}$	$19.6°\ 2^{\pi}$
Maximum angular tracking speed	0	$\lvert\beta - G\rvert \leq 233$ µrad $\lvert\beta - L\rvert \leq 476$ µrad	$\lvert\beta - G\rvert \leq 288$ µrad $\lvert\beta - L\rvert \leq 388$ µrad
Maximum lead angle	35 µrad	72 µrad	70 µrad
Specify uncertainty range	±0.1°	±0.1° ±0.15°	±0.1° ±0.15°

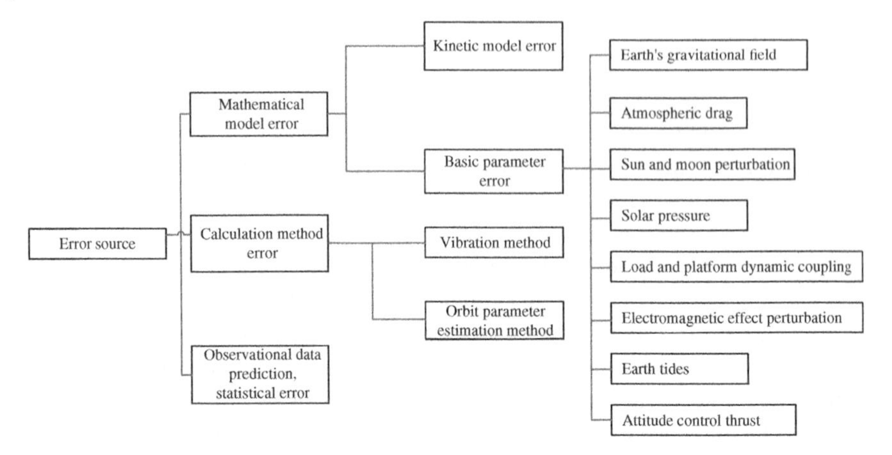

Figure 6.3 The error of precise orbit positioning.

require real-time high-precision orbit data and attitude control data during acquisition and high-precision tracking. There are two ways to obtain orbit data: the method of preinjection requires high-precision orbit determination on the ground. The initial value is extrapolated on the ground and the real-time calculation method on the satellite using the high-precision model requires high computing power and methods on the satellite. In addition, the prediction and calculation accuracy of satellite orbits are also affected by uncertain factors such as atmospheric model errors (the error can reach 30–40% of the predicted value), changes in the surface/mass ratio (the light pressure in different directions and the influence of the atmosphere are different), and other influence of uncertain factors. The precision orbit determination error is shown in Figure 6.3.

6.3.2 Satellite Attitude

Attitude sensors used in attitude control of satellite platforms mainly include infrared horizon, solar sensor, gyroscope based on inertia, accelerometer (inertial), star sensor (optical), in addition to magnetometer, land mark sensor, etc. With the application of star sensors on satellites, the accuracy of satellite attitude measurement and control has been greatly improved [13].

The main performance parameters of GEO platform and LEO platform at home and abroad are shown in Tables 6.2 and 6.3, respectively. It can be seen from the data in the table that the attitude control performance provided by the satellite platform in China is comparable to the advanced level abroad.

Table 6.2 Main performance parameters of GEO platform related to laser communication.

Index	Chinese communication satellite	Artemis	ETS-VI
Roll, pitch	±0.06°	0.03–0.06°	±0.05°
Yaw	±0.2°	0.08–0.24°	±0.15°
Attitude stability	10^{-4}°		
Position retention accuracy	±0.05°	±0.07°	

Table 6.3 Main related performance parameters of LEO platform related to laser communication.

Index	Chinese sun synchronization platform	Landsat-4	SPOT-4
Attitude determination accuracy	0.03 (star sensor) 0.15 (no star sensor)	0.03°	0.03°
Attitude stability	10^{-3}–5×10^{-4}°/s	10^{-6}°/s	10^{-4}°/s

6.3.3 Uncertain Angle of Pre-Cover

The biggest feature of the inter-satellite link is that the link is dynamic. A communication link between two satellites is established as needed, and the link is released after the communication task is completed. Before each link is established, the two communication terminals that establish the link need to be able to point to each other accurately. However, due to the existence of various uncertain factors, there is an error in the angle of the communication terminal pointing to the opposite party, and the opposite terminal cannot accurately appear in a predetermined position, but can only appear in an area with a certain probability. In order to establish contact with each other, due to the mutual positional relationship and pointing error between the two terminals, it is necessary to scan and search an area [14, 15].

There are many factors that affect the preset uncertainty angle, including the attitude accuracy of the satellite where the terminal is located, the orbit prediction accuracy of the satellite, the satellite orbit perturbation, the execution accuracy of the terminal pointing mechanism, etc. Among them, the attitude accuracy and orbit prediction accuracy play a decisive role. If the value of the preset uncertainty angle is large, a larger scanning area will be required, which will seriously restrict the realization of the acquisition time and acquisition probability index, and will

Table 6.4 The error distribution of the OICETS satellite at open loop state.

Error source	Random quantity (°)	Deviation (°)
Attitude determination error	0.002	0.1412
Attitude compensation residual error	0.0104	
Attitude measurement equipment installation accuracy		0.0316
Orbit calculation error	0.018	0.044
Execution time precision		0.037
Coarse pointing mechanism pointing accuracy	0.0212	
Coarse pointing machine assembly accuracy		0.0309
Optical assembly accuracy		0.0225
Total	0.03	0.160
Sum	0.190	

also require a larger beacon optical power, resulting in an increase in power consumption and volume, increasing the difficulty of implementation [16, 17].

The open-loop alignment error distribution of the OICETS satellite carrying the LUCE laser communication system is shown in Table 6.4.

6.3.4 Satellite Vibration Problem

In the space optical communication link system, due to the narrow beam, the interference caused by the vibration of the platform may be significantly larger than the allowable value assigned by the laser communication system PAT, and measures need to be taken to suppress it. According to the platform vibration spectral density, the platform can be divided into three categories: stable platform (platform vibration spectral density <2 grad), normal platform (platform vibration spectral density <20 grad), and poor platform (platform vibration spectral density >20 grad). Two types of normal platforms and poor platforms should consider isolation measures, including passive vibration isolation measures and active vibration suppression measures. Passive vibration isolation measures can reduce vibration by 10–30%, and active vibration suppression measures can reduce vibration by 1/10. Table 6.5 lists the vibrational spectral densities of typical platforms.

From the typical platform vibration spectral density above, it can be seen that the vibration spectrum is mainly concentrated in the 1–100 Hz range. Therefore, it is required that the vibration suppression spectrum of the system should reach several hundreds of hertz, that is, the servo bandwidth of the tracking system should be several hundred hertz. More importantly, in the capture process,

Table 6.5 The vibration spectrum density of some typical platforms.

Spacecraft	Vibration spectrum distribution			
	1–10 Hz	10–100 Hz	100–1 kHz	1–10 kHz
ASTRO-SPAR (3145 kg)	0.58	0.45	0	0.59
STRV-2	0.22	0.03	0.02	0.22
RME (1040 kg)	0.10	0.35	0.17	0.40
CASSNI (2175 kg)	1.33	1.48	0.01	1.92
Olympus (2000 kg)	7.76	4.01	1.20	11.49
Motoroal	17.56	4.43	3.93	18.49
Landsat	3.42	6.07	3.21	7.62
Bosch	61.7	13.18	6.29	63.22
Hrdls	18.37	18.79	7.66	26.97
Iridium	4.52	99.19	76.34	124.99
Shuttle	219.65	477.95	15.15	519.29

because the servo system has not yet closed the loop, passive methods such as isolation and other special methods must be used to reduce vibration, or the frame rate of the captured data must be increased to overcome this effect.

6.3.5 Dynamic Coupling Problem

In the satellite laser communication system, the movement of other loads or components of the satellite is another important reason for the shaking of the base platform by combining with the multi-body and flexible dynamics of the star. For ordinary satellites, this kind of jitter mainly comes from the movement (vibration) of the solar panel, the movement of the onboard rotating antenna, the operation of the flywheel of the satellite attitude control system, and the sloshing of the liquid in the storage tank. The preliminary simulation analysis of the power failure of the laser communication terminal applied to different common satellite platforms shows that the orbit maintenance adjustment has a great influence on the platform, and the orbit adjustment should be avoided as much as possible during the communication process [12–14].

6.3.6 Influence of Background Stray Light

Optical communication terminals in space will inevitably receive background radiation. Background radiation is processed together with the wanted signal, resulting in a degraded overall system performance. Background radiation is one of the main reasons for the decrease of the sensitivity of the optical terminal.

Excessive background radiation may also cause saturation of the photoelectric sensor, or even burn out [9].

Background light sources mainly include solar radiation, the Earth's reflection of the solar spectrum, and the radiation of planets and stars. It can be seen from the order of magnitude of various light sources that the influence of solar spectral irradiance and Earth irradiance spectrum on the spaceborne optical receiver is 8–10 orders of magnitude greater than that of other light sources. Therefore, the influence of the solar spectrum and the irradiance of the Earth should be mainly considered.

For satellite platforms in different orbits (such as LEO, MEO, and GEO), the effect of background light is different when they optically communicate with each other. For LEO–MEO and LEO–GEO optical communication links, MEO, GEO terminal receiving antennas will take the Earth's reflection and scattering of sunlight as the main background. Assuming that the heights of MEO and GEO from the Earth are 10,000 and 36,000 km, the field of view coverage of the Earth to MEO and GEO is about 20° and 13°. Since the field of view of the optical receiving system on the actual MEO and GEO is usually several mrad and dozens of μrad, the Earth background will fill the field of view of the MEO and GEO receiver, which is an extended radiation source. The reflection of sunlight by the Earth is related to the angle between the sun, Earth, and GEO, as shown in Figure 6.4. When the included angle is zero, the background noise generated by the solar spectrum reflected by the Earth to GEO is the strongest.

The direction of the optical communication terminal on the LEO varies with the location, and the influence of the background light on the LEO mainly comes from the illumination of the solar spectrum.

Since there are no particles in space to scatter sunlight, the pointing azimuth of the LEO receiver, the angle between the line connecting LEO and the sun, and the

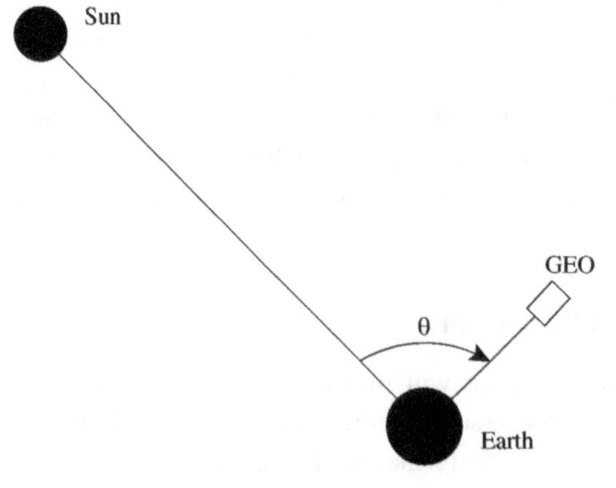

Figure 6.4 The schematic drawing of the relative position in the midst of the sun, the Earth, and GEO.

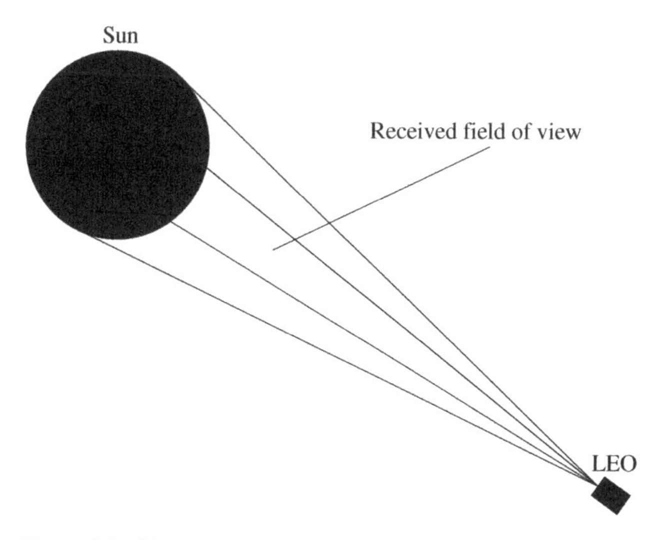

Figure 6.5 The schematic drawing of the impact of the sun spectra on LEO.

receiving field of view of the receiver will be important factors that determine how much LEO is affected by background stray light. As shown in Figure 6.5, if the sun is included in the field of view of the receiver, there will be strong background stray light, and vice versa.

The background light power reflected by the Earth received by the receiver can be calculated by the following formula, that is, the light power generated by the photons of the background radiation received by the detector on the detector is:

$$P_r = W(\lambda)\Delta\lambda\Omega_r A_r \tag{6.1}$$

where:

$W(\lambda)$: background radiation spectral function
Ω_r: receiver field of view (s_r)
$\Delta\lambda$: bandwidth of light (W_n)
A_r: the optical aperture area of the receiver.

At two wavelengths:

$$W(800\ \text{nm}) = 0.1\ \text{w/cm}^2 \cdot \mu\text{m} \tag{6.2}$$

$$W(1550\ \text{nm}) = 0.04\ \text{w/cm}^2 \cdot \mu\text{m} \tag{6.3}$$

Considering the Doppler effect drift of 0.3 nm, the temperature and chirp drift of the distributed feedback laser can be controlled at 0.8 nm, and the temperature and chirp drift of other lasers can be controlled at 1 nm, and a narrow-band filter with a bandwidth of 3 nm can be used. If the aperture of the receiving antenna is set to be 200 nm and the receiving field of view is 22 μrad, the power of the Earth's background light on the detector is

$$P_r(800\,\text{nm}) = 3.77 \times 10^{-12} W \tag{6.4}$$

$$P_r(1550\,\text{nm}) = 1.13 \times 10^{-12} W \tag{6.5}$$

The noise power generated by the background light on the detector is below the order of $10^{-20}W$, far less than the noise of the detector itself, indicating that it is feasible to use a narrow-band filter with a bandwidth of 3 nm. In the case of direct sunlight, the field of view of the background light is 22 μrad, then the power of the background light shining on the detector is

$$P_r(800\,\text{nm}) = 4.56 \times 10^{-7} W \tag{6.6}$$

$$P_r(1550\,\text{nm}) = 1.82 \times 10^{-7} W \tag{6.7}$$

In the case of direct sunlight, the received background light is much larger than the signal light power, and it is very difficult to achieve tracking and communication. Therefore, direct sunlight should be avoided when making communication requirements.

6.4 Transmitter Design

6.4.1 Choice of Laser

In inter-satellite link applications, the most promising lasers are mainly semiconductor lasers. The strip-shaped double-heterostructure semiconductor laser is currently the best choice, and AlGaAs/GaAs lasers are used in the SILEX and ETS-VI missions. The typical size of the laser chip is 0.1 × 0.2 × 0.2 mm, which is fixed on the copper heat sink. The heat sink itself is fixed on the thermoelectric temperature control device. The chip consists of n-type and p-type GaAs and AlGaAs grown on a GaAs substrate. The current is injected into the active layer, and when the current value exceeds the threshold level, the laser will be emitted from the end face, and the conversion efficiency is as high as 34%. The cross section of the laser beam is elliptical, and the typical size at the end face is 1 × 4 μm, so it is highly divergent. This requires a short focal collimating lens to convert it into parallel light. There is also a multi-lens assembly to correct for spherical aberration, coma, and astigmatism of the laser wavefront. The wavelength spectrum of the emitted laser light is discrete modes spaced 0.4 nm apart. When the laser is directly modulated by the excitation current, several modes oscillate simultaneously. Several devices are available [10–14]:

1) The GaAs/AlGaAs laser can efficiently emit light in the wavelength range of 0.80–0.86 μm, and the output power of the laser is expected to exceed 500 mW in the near future. The current output power of such lasers is around 30 mW and has a lifespan of about $10^5 - 10^6$ hours (under terrestrial conditions). Intelsat has carried out a lifetime test on the wide-stripe laser at a high temperature of 120 Mbit/s PPM, peak output power of 150 mW, and 70 °C. From the test results, it is inferred that the lifetime of the laser at 25 °C is about 50 years. According to Spectra Diode Laboratories, their single-mode 100 mw continuous output power device has an inferred lifetime of more than 60,000 hours at 25 °C.
2) Nd:YAG: This is another very interesting laser that can be used for space communications. Its core is a Chin-doped YAG rod, which can emit high-power lasers in the 1.06 μm spectral band, as well as 1.3 μm and 0.53 pm (frequency doubling output) lasers. Optically pumped YAG rods excited by high-power semiconductor lasers have greatly improved in efficiency and reliability compared to flashlamp-excited lasers. However, its application in space is still subject to many limitations, mainly in the relatively low efficiency (about 8%), the limited range of wavelength reuse, and the need for external modulators that can work at high power.

The selection of the laser should take into account factors such as the directionality, monochromaticity, and brightness of the laser light emission under the premise of the given power and volume requirements of the system. In theory, gas lasers are the best in the above three performances, followed by solid-state lasers, and semiconductor lasers are the worst. However, from the perspective of engineering and practicality, especially in special environments such as space applications, solid-state lasers are superior to gas lasers in terms of reliability, life, weight, and power consumption. At present, the semiconductor lasers used in space are all products of American Spectrum Corporation, and domestic products cannot reach the level of space application in the near future. Based on the above considerations, the 30 mW GaAlAs semiconductor laser of American Spectrum Corporation was selected.

6.4.2 Wavelength Selection

When choosing a wavelength, attention should also be paid to satisfying the wavelength ratio theorem in electromagnetic theory, that is, in theory, the devices and equipment used for electromagnetic wave emission, transmission, and reception are proportional to the wavelength. As will be seen below, the gain of the transmitter has nothing to do with the wavelength, so that the wavelength can be selected relatively short on the premise of ensuring the gain of the transmitter, so as to reduce the volume and weight of the entire transmitter. But in the special environment of inter-satellite communication, the main contradiction is to choose a suitable laser transmitter. Since the 30 mW GaAlAs semiconductor laser has been selected as the laser transmitter in the first step of the design process, and its wavelength is 830 nm, the wavelength of the transmitter has actually been determined to be 830 nm.

6.4.3 Selection of the Diameter of the Transmitting Antenna

The principle that should be followed in selecting the aperture of the transmitting antenna is to select an appropriate aperture value under the constraints of the entire system to maximize the gain of the transmitter. Since the optical communication antenna of the two satellites needs to extend out of the cabin during the optical communication process, there is a problem of mutual occlusion between it and other parts of the satellite, so the size of the antenna aperture is strictly limited. In addition, in the process of inter-satellite communication, the laser beam used for communication is sent to the terminal on another satellite that communicates with it through the transmitting antenna (actually an optical telescope), that is to say, the telescope needs to emit laser beams to the laser beam imaging. Since the limit distance between two optical communication terminals between satellites is 45,000 km, there is a strict limit on the divergence angle of the communication laser beam (in the order of microradians), which puts forward extremely strict requirements on the image quality of the telescope. Therefore, the actual manufacturing level and processing technology level of optical materials must be considered when choosing a telescope [12, 16].

6.4.4 Calculation of Transmitting Antenna Gain

The link budget of the optical inter-satellite link can adopt the same system as that of the microwave system, but it must be noted that the gain of the optical antenna has a great influence on the design of the whole system. First, the antenna gain G_T can be expressed as

$$G_T = \left(\frac{\pi D_T}{\lambda}\right)^2 \eta_T \tag{6.8}$$

where η is the antenna efficiency, generally taken as 0.6–0.75, expressed in decibels as

$$[G_T] = 10LG\left(\frac{\pi D_T}{\lambda}\right)^2 \eta_T = 120.9\,db \tag{6.9}$$

6.5 Receiver Design

6.5.1 Selection of Receiver Detector

A good optical communication system needs a good optical receiver, and the photodetector is the key component of the optical receiver. For this purpose, the requirements for the detector are: (i) corresponding to the light wave of the wavelength used, it must have high sensitivity (fast response to weak light signal pulses); (ii) it must have a wide enough bandwidth to accommodate the received light signal bandwidth; (iii) in the process of receiving and demodulation, the additional noise should be small.

The following are several types of light detectors that can be used for space communications:

1) Charge-coupled devices (Charge-Coupled-Devices, referred to as CCD). As a new type of photoelectric conversion device, CCD sensor has been widely used in aerospace, aviation, and other fields. In the satellite laser communication tracking and aiming system, the CCD can be used as a position sensor to detect the position of the laser spot, and further obtain the direction of the laser beam emitted by the other satellite. Since each pixel of the CCD corresponds to an orientation of the light beam, by selecting the size of the CCD target surface, the number of pixels and the corresponding optical path, the requirements of the field of view and accuracy can be taken into account at the same time. CCD is a charge accumulation-type device, and its readout circuit is relatively complex, and the readout speed is also limited to a certain extent. Generally, the frame frequency is up to several kilohertz. However, its high resolution, large dynamic range, and no dead zone have made CCDs also used as precision aiming sensors.
2) Four-quadrant detector. The four-quadrant detector is divided into four quadrants by a cross channel on one photosensitive surface. Each quadrant is equivalent to a photocell. When the laser is incident vertically, the focused spot illuminates the center of the four quadrants. The quadrant receives the same light intensity, and the output photocurrent is also the same. When the laser beam is incident at other angles, the position of the light spot is also shifted, and the photocurrent's output by the four detectors are also different. By performing differential processing on the current's output by the four detectors. The error signal of the light spot off center can be obtained. The light sensitivity of the four-quadrant detector is very high, the detection frequency can reach several kilohertz, the peripheral circuit is relatively simple, and the sensitive band reaches 1550 mn, so it is an ideal choice for precision aiming sensors. However, the photosensitive surface of the four-quadrant sensor is relatively

small, and there is a dead zone. In addition, since the four-quadrant sensor actually only detects the light intensity of four areas, the accuracy and linearity of the calculated position shift are affected by the size and shape of the light spot.

3) APDs. The communication receiver must have the highest sensitivity and relatively low dynamic range at the corresponding wavelength, and the dynamic range of the inter-orbit link from LEO to GEO is generally about 6 dB. In this regard, the performance of silicon APDs in the optical fiber communication market is the most prominent. Silicon APDs are particularly sensitive because of their very high electron-to-hole ionization ratio. RCA's silicon avalanche photodiode device has a photoelectric conversion efficiency of more than 90% near 0.8 μm, and a life span of more than 10 years.

The selection of the receiving detector is very important to improve the sensitivity of the optical receiver. To effectively convert the luminous flux of the receiver into signal power, the noise generated by the optical receiver should be as small as possible. In addition, the noise generated when receiving light is thermal noise of the light receiving system and shot noise caused by various currents. Considering sensitivity and noise factors, and matching the emission wavelength, APD is the best choice. ESA has successfully completed space experiments in the SILEX system used in the APD detector. The SLIK APD produced by EG&G Company is used here, and its sensitivity is −59 dBm.

6.5.2 Selection of Receiving Antenna Aperture

When selecting the aperture of the receiving antenna, in addition to factors similar to the selection of the aperture of the transmitting antenna, the relationship between it and the gain of the receiving antenna should also be considered. As will be seen in the discussion below, the larger the receive antenna aperture, the greater the gain. Here, the diameter of the receiving antenna is selected as 20 cm.

6.5.3 Calculation of Receiving Antenna Gain

Assuming that the receiver is in the far field relative to the transmitter, then what the receive antenna will receive will be a plane wavefront. The reduction in receive antenna gain due to misalignment of the receive antenna can then simply be given by the projection of the incident aperture on the incident wavefront:

$$G_R = \left(\frac{\pi D_R}{\lambda}\right)^2 \eta_R \cos\varepsilon \tag{6.10}$$

Usually the misalignment angle ε is small, the cosine term is always close to 1, so the gain of the receiving antenna is

$$G_R = \left(\frac{\pi D_R}{\lambda}\right)^2 \eta_R \tag{6.11}$$

Expressed in decibels as

$$[G_R] = 10\log\left(\frac{\pi \times 0.2}{830 \times 10^{-9}}\right)^2 0.7 = 116\,\text{dB} \tag{6.12}$$

6.5.4 Calculation of Received Power

After completing the design of the transmitter and receiver, the received power of the system has actually been determined, which can be described as:

$$P_R = P_T G_T T_T \left(\frac{\pi D_R}{\lambda}\right)^2 G_R T_R$$

where:

R: distance between transmitter and receiver
T_T, T_R: transmit system and receive system losses.

References

1 Chan, S. (2005). *Architecture for a Space-Based Information Network with Shared On-Orbit Processing*, 154–170. Massachusetts Institute of Technology.

2 Quintas, D. and Friderikos, V. (2012). All optimal solutions in STDMA scheduling. *20th European Signal Processing Conference*. Bucharest, 834–838.

3 http://www.ccsds.com, 2014-09.

4 Ollie, L., Larry, B., Art, G. et al. (2005). GPS III system operations concepts. *IEEE Aerospace and Electronic Systems Magazine* 20 (1): 10–18.

5 CCSDS 714.0-B-2 Space communication protocol specification (SCPS) – transport protocol.

6 CCSDS 133.0-B-1 Space packet protocol.

7 Space Communication Architecture Working Group (2006). *NASA Space Communication and Navigation Architecture Recommendations for 2005–2030*, 18–22. NASA.

8 Chang, S., Kim, B.W., Lee, C.G. et al. (1998). FSA-based link assignment and routing in low-earth orbit satellite networks. *IEEE Transactions on Vehicular Technology* 47 (3): 1037–1048.

9 Knibbe, T.E. (2016). Spatial tracking using an electro-optic nutator and a single-mode optical fiber. *SPIE* 35 (16): 309–317.

10 Komukai, T. (2014). Performance evaluation of laser communication equipment. *SPIE* 36 (92): 41–50.
11 Roberts, W.T. and Wright, M.W. (2013). The lunar laser OCTL terminal (LLOT) optical systems. *SPIELASE* 32 (43): 11–19.
12 Tolker-Nielsen, T. and Oppenhaeuser, G. (2015). In-orbit test result of an operational optical inter satellite link between ARTEMIS and SPOT4, SILEX. *SPIE* 46 (35): 1–15.
13 Sodnik, Z., Lutz, H., and Furch, B. (2016). Optical satellite communications in Europe. *Proceedings of SPIE the International Society for Optical Engineering* 75 (87): 87–97.
14 Pease, R. (2015). Optical laser communication system scar venichein metromarkets. *Light Wave* 45 (9): 23–27.
15 Komukai, S. (2014). Performance evaluation of laser communication equipment onboard the ETS_VI satellite. *SPIE* 36 (92): 41–50.
16 Arnon, S. (2016). Optimum transmitter optics aperture for free space satellite optical communication. *SPIE* 28 (11): 252–264.
17 Kenichi, Janhaiku, and Saiwaiku (2017). Experimental operations of laser communication equipment on board ETS_VI satellite. *SPIE* 29 (90): 264–275.

7

Inter-Satellite Laser Capture, Aiming, and Tracking System

7.1 Introduction

As the name suggests, the basic steps of the APT system can be divided into three stages: acquisition (Acquisition, A), alignment (Pointing, P), and tracking (Tracking, T). It is required that after the system completes the capture, the beacon light can be successfully received by the camera monitoring system; after the system is aligned, the signal light can be successfully received by the signal detector receiving system, and tracking must ensure point-to-point locking. Establish and maintain the normal working state of the communication link between two points.

The basic block diagram of the space laser communication APT system is shown in Figure 7.1.

APT technology of space optical communication is a complex technology integrating light, machine, and electricity. It can be seen from Figure 7.1 that it can be roughly divided into the following parts: optical platform system, coarse tracking system, advanced aiming mechanism, fine tracking system, information target light emission, and CCD detector. The functions of the main components are as follows.

As an important part of atmospheric laser communication, the optical antenna system is actually an optical telescope system, which is very worthy of our attention and research [1].

Generally, the basic requirements for optical antennas are high optical quality, low occlusion rate, small thermal expansion coefficient of optical antenna materials, high mechanical strength, large aperture, high magnification, near-diffraction, etc.

The optical antenna system includes a transmit antenna and a receive antenna. The quality of optical antenna performance directly affects the reliability of communication. The transmitting and receiving antennas of the atmospheric laser communication system use the lens system. Optical antennas can generally be classified as listed in Table 7.1.

Laser Inter-Satellite Links Technology, First Edition. Jianjun Zhang and Jing Li.

Published 2023 by John Wiley & Sons, Inc.

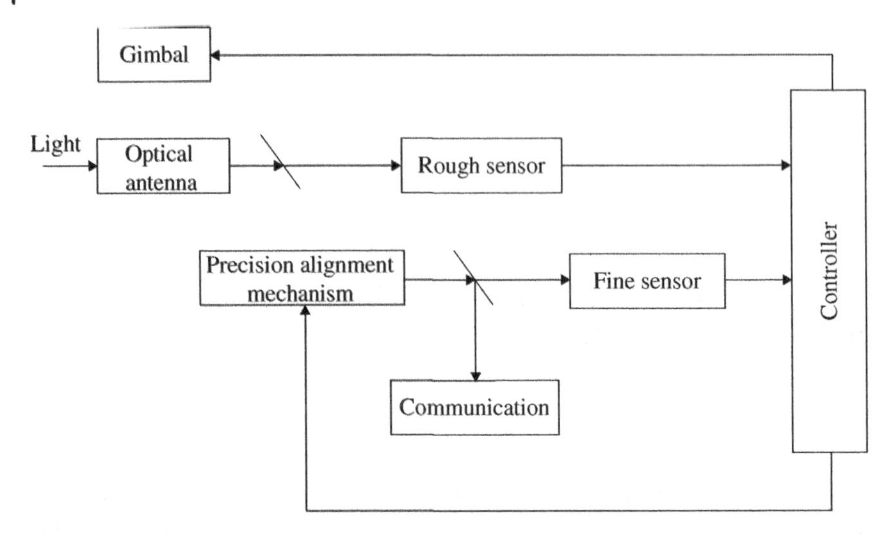

Figure 7.1 APT system basic schematic.

Table 7.1 Comparison of transmissive and reflective telescopes.

	Advantage	Disadvantage
Transmissive telescope	There is no occlusion on the receiving and light center, it is easier to process spherical mirrors, and aberrations can be eliminated through optical design	The loss of light energy is large, the installation and debugging are difficult, and the actual application is less
Reflecting telescope	Low material requirements, light weight, small light energy loss, and no chromatic aberration	It is difficult to meet the imaging requirements of large aperture and large field of view due to occlusion of the center of the receiving light

A few concepts are briefly introduced below: capture area, alignment area, and tracking area. These three areas have a certain contain and be contained relationship, as shown in Figure 7.2.

✓ Capture area: also called uncertain area, which refers to the area where the transmitter is a priori set. The range is large, and for long-distance communication, the approximate location of the uncertain area can be determined with the

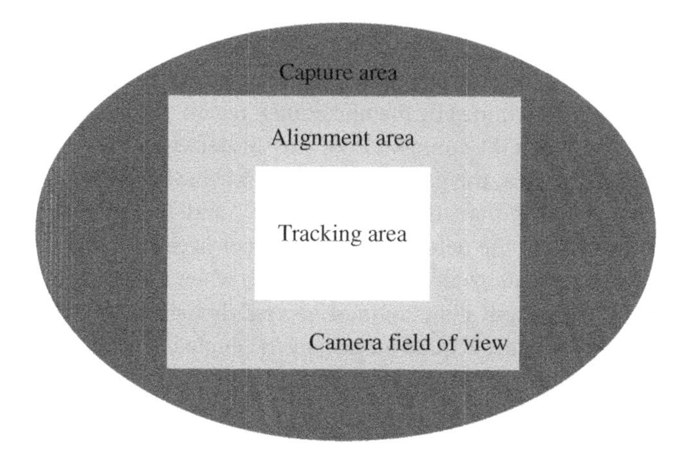

Figure 7.2 The relationship structure diagram of scanning area, alignment area, and tracking area.

help of the GPS system. For the point-to-point wireless laser communication system on the ground, the determination of the captured uncertain area can also be completed with the help of the electronic theodolite system.

✓ Aiming area: the area where the beacon light enters the field of view (FOV) of the camera monitoring system. The alignment area must be inside the uncertainty area and include the tracking area.

✓ Tracking area: the signal light enters the area where the signal detector can respond. The range is small and the establishment and maintenance of the signal link must be guaranteed in this area. Generally, when the hardware structure is fixed, it is necessary to ensure that the communication link can be successfully established when the light spot enters a certain area in the center of the camera's monitoring FOV.

7.2 Acquisition

Capture is the process of going from the capture area to the alignment area. For long-distance APT, the first step is to determine the location of the uncertain area with the aid of auxiliary tools. The main methods are GPS positioning and electronic theodolite positioning. To capture is to capture the beacon light within an indeterminate area. The captured tracking mechanism is closed loop control. The selection of capture scheme and the selection of capture path are the main tasks of capture unit design [2, 3].

7.2.1 Capture Scheme

The main capture schemes are stare-gaze, stare-scan, and scan-scan. The choice of the specific capture scheme is determined by the uncertainty region, the emission FOV and the detection FOV. When the emission FOV and the detection FOV can completely cover the uncertain area, the gaze-gaze capture scheme can be used; when the emission FOV is larger than the detection FOV, and the emission FOV is within the uncertain area, the detection FOV may not necessarily be in the uncertain area, and staring is adopted. Scanning scheme: when the emission FOV and the detection FOV are equivalent, and not necessarily both within the uncertain area, the scanning-scanning acquisition scheme is adopted. It is obvious that the time consumption of these three capture schemes increases in turn, and the capture probability decreases in turn.

The gaze-scan and scan-scan capture schemes are discussed in detail below. For the comprehensive consideration of the system in this paper, the capture scheme in the laboratory is gaze scanning. The acquisition scheme adopted by the final long-distance wireless laser communication system is scan-scan [3].

7.2.1.1 Stare-Scan

Gaze – When scanning, the receiver always keeps staring, that is, without any attitude adjustment; the sender scans with a specific scanning path in the uncertain area. The capture method is suitable for the situation where the beam divergence angle (transmission FOV) of the signal light beam at the transmitting end is larger than the uncertainty area and the receiving FOV (detection FOV) of the optical antenna at the receiving end is smaller than or equal to the uncertainty area. At this point, there is another point that needs special attention: to ensure the reliability of scanning. That is to say, there should be a part of the overlapping area between the scanning point and its adjacent scanning points. The size of the overlapping area is determined by the coverage factor ε_t to guarantee not to miss the uncertain area. The capture time of the scan-gaze method t_{acq} can be expressed as:

$$t_{acq} = \left(\frac{\theta_{unc}^2}{\theta_{unc}^2}\right) \times t_s \times N_s \tag{7.1}$$

ε_t is coverage factor and $\varepsilon_t = (1-k)^2$, where $k = 10 - 15\%$;
θ_{unc} is the opening angle between the uncertainty region and the emission point;
θ_{laser} is the $1/e$ width of the laser beam;
t_s is dwell time for the scan point;
N_s is the number of scan points.

The capture probability p_{acp} can be expressed as:

$$p_{acp} = p_{unc} \times p_{det}$$

p_{unc} is the coverage probability of the target for the uncertain area;
p_{det} is the detection probability of the receiver.

7.2.1.2 Scan-Scan

During scanning-scanning, both the receiving end and the detecting end scan with a specific scanning path in the uncertain area. In most cases, the skip-scanning scheme is adopted. This method is suitable for the situation that the beam divergence angle (transmission FOV) of the signal light beam at the transmitting end and the receiving FOV (detection FOV) of the optical antenna at the receiving end are both smaller than the uncertain area. The reliability of scanning should also be guaranteed. Since the ratio of the uncertainty area to the FOV received by the detector is greater than 1, so by this method, captured time t_{acq} is longer:

$$t_{acq} = \left(\frac{\theta_{unc}^2}{\theta_{laser}^2 \times \varepsilon_t}\right) \times \varepsilon_t \times N_s \times \left(\frac{\theta_{unc}^2}{\theta_{fov}^2 \times \varepsilon_t}\right) \times t_{rs} \times N_{rs} \tag{7.2}$$

ε_t is coverage factor and $\varepsilon_t = (1-k)^2$, where $k = 10 - 15\%$;
θ_{unc} is the opening angle between the uncertainty region and the emission point;
θ_{laser} is the $1/e$ width of the laser beam;
t_s is dwell time for a single scan point of the sending terminal;
N_s is the number of scanning points for the sending terminal;
θ_{fov} is the FOV of the detection terminal;
t_{rs} is pause time for a single scan point on the detection side;
N_{rs} is the number of scanning points for the FOV angle of the detection end.

After determining the approximate azimuth of the uncertain area, the two terminals of the wireless laser communication system (represented by station A and station B in this paper) scan in their respective uncertain areas, and the detection FOV of station A. Within the divergence angle of the beacon light of station B, and after adjustment, the main axis of the beacon light and the main axis of the detection FOV can be coaxial, as shown in Figure 7.3.

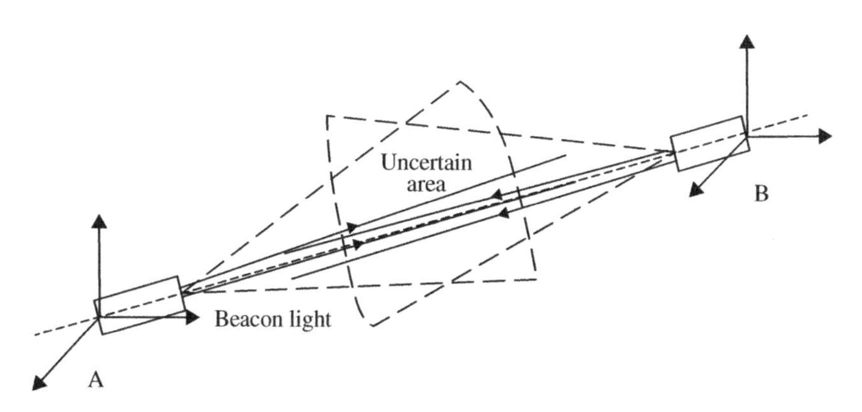

Figure 7.3 The schematic diagram of scanning-scanning.

When capturing, stations A and B are powered on, and both stations A and B must scan in the uncertain area with their specific capture paths. At the same time, the receiving link is in an open state to detect in real time whether the beacon light of the other party enters the respective monitoring FOV. Among them, the scanning methods of station B and station A are different. The scanning method of station A is relatively continuous, and the dwell time interval between scanning points is short. Compared with the continuous scan of station A, the scan method of station B should be called skip scan, and the dwell time between scans is longer. That is, when station A does not complete a scanning action, station B stares; then station B jumps and points to another direction, and station A performs the next scanning action at this time. In any scanning process, stations A and B must each ensure that there is an overlapping area between their adjacent scanning points to ensure that the probability of capture must exist, but it is only a matter of the length of capture time, and it is not possible to miss the area where a certain signal light may exist [4].

7.2.2 Capture Path

After entering the uncertain area, the rough scanning capture path mainly includes raster rectangular scanning (as shown in Figure 7.4a), helical scanning (as shown in Figure 7.4b), and rectangular helical scanning (as shown in Figure 7.4b and c). There are many other scanning methods: rose-shaped scanning, Lissajous scanning, etc., but due to the limitation of the dimensionality of the mechanical structure, the system mostly adopts the three methods shown in Figure 7.4.

Raster scanning: For mechanical structures, this scanning method is simpler and easier to implement; however, this scanning method requires a long acquisition time, and the scanning path cannot be scanned according to the probability of occurrence of optical signals from high to low. Helical scanning: This scanning method can effectively cover the uncertain area, and there will be missed scanning in the uncertain area, and only one third of the area of the uncertain area can be searched. Rectangular helical scanning: It is an improvement over the first two scanning methods. It combines the advantages of raster scanning and helical scanning. Scan in uncertain areas. In engineering practice, the capture time and capture probability need to be balanced in the design. To reduce the average capture time, we must sacrifice the capture probability; to increase the capture probability, the average capture time must be sacrificed. How to find a balance between capture time and capture probability is also a major problem in capture design [5, 6].

Starting from the relevant characteristics of wireless laser communication, scanning should be adopted in the acquisition method in long-distance communication. Scanning, specifically, skipping. Scanning mode: the scanning path is a rectangular spiral scan. After the system completes the acquisition process, the

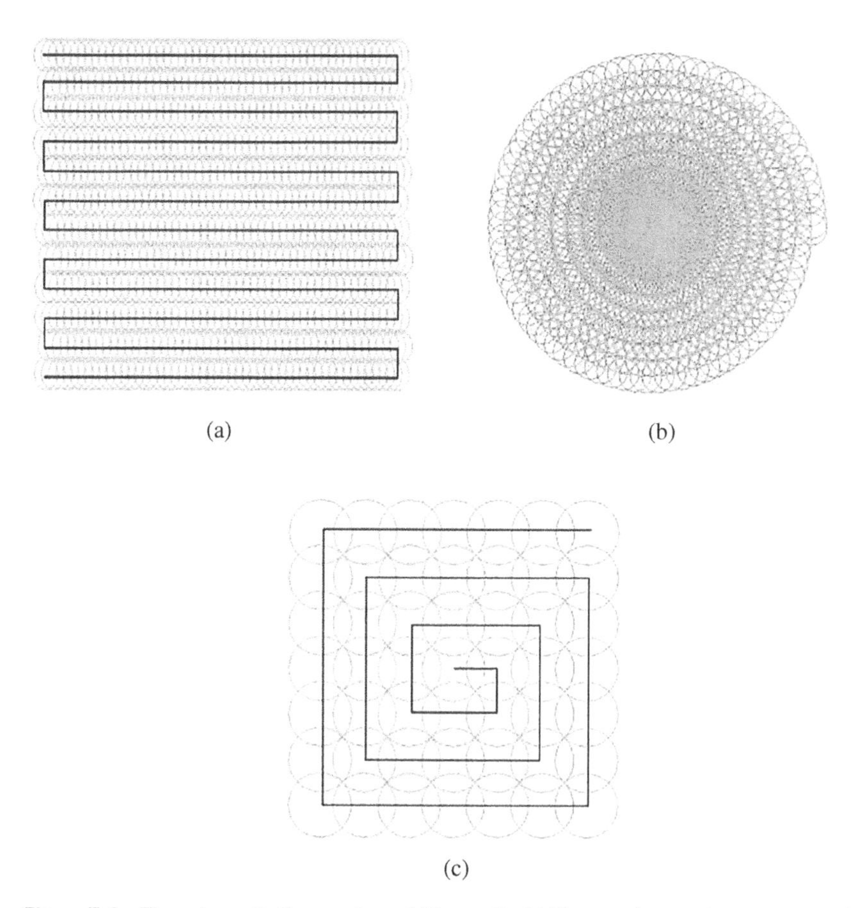

Figure 7.4 The schematic figure of acquisition path. (a) The rough scanning capture path; (b) helical scanning; (c) rectangular helical scanning.

beacon light can be successfully received by the tracking detector. The next step is to complete the alignment process, that is, to move the beacon light to the center of the FOV, so that the signal light can be received by the signal detector.

7.3 Pointing

Alignment is the process of introducing optical signals from the alignment area into the tracking area. After capturing the target, the system enters the alignment phase. At this time, the system is in a closed-loop working state, and the tracking servo system is controlled according to the spot off-target amount provided by the detector, and the azimuth and pitch angles of the optical antenna are adjusted to make the received beacon spot close to the center of the coarse tracking FOV. When the position of the light spot enters the predetermined fine tracking area, the fine tracking loop is started, but the coarse tracking control does not stop all the time. If the two terminals of the wireless laser communication system are still unable to communicate at this time, the small-step displacement is randomly performed in the fine tracking area until the communication link between the two parties is established [7].

7.4 Tracking

After the communication link is established, the system is also in a closed-loop working state, and the position of the signal light on the photosensitive surface of the camera at this time is recorded, and this position becomes the relative zero position. When the signal light is offset, use the specific information of the offset to control the servo to position the signal light at this position again [8].

7.4.1 Analysis of Tracking System Beacon Beam Divergence

In the working process of the tracking system, the divergence angle of the beacon beam is directly related to the performance of the entire tracking system. Figure 7.5 is a demonstration diagram of the laser communication link.

The optical power received by the tracking detector can be expressed as:

$$P_r = P_t \times L_{geo} \times L_{atm} \times \eta_t \times \eta_r \tag{7.3}$$

Among them, P_r indicates the optical power received by the tracking detector, P_t indicates the transmitting power of the laser at the transmitter, L_{geo} represents the

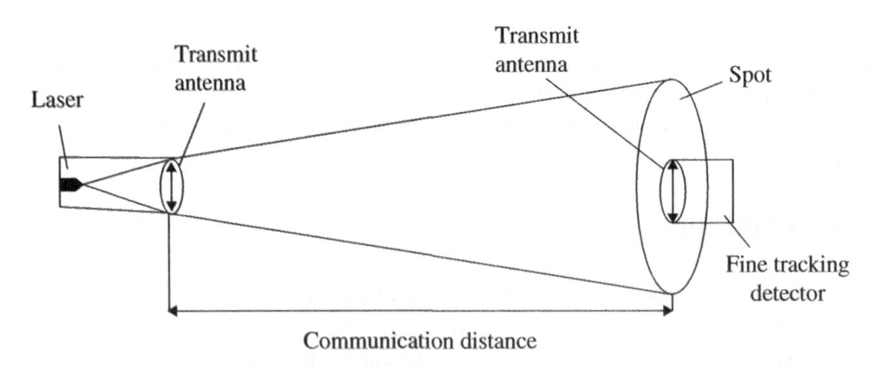

Figure 7.5 Laser communication link diagram.

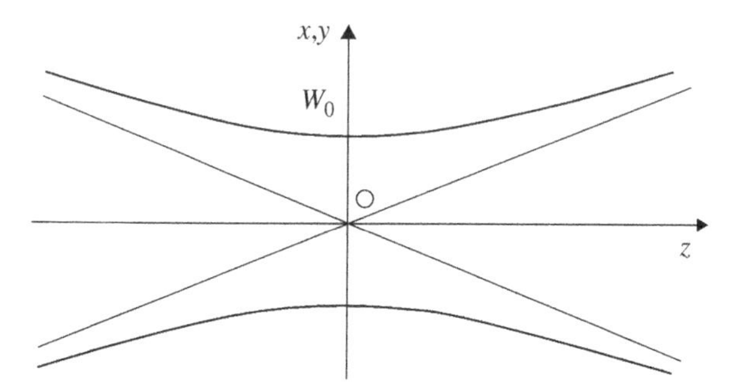

Figure 7.6 Gauss beam propagation characteristics.

geometric error at the receiver, $L_{atm} = e^{-\alpha L}$ represents absorption and scattering attenuation due to turbulent atmosphere, α is the atmospheric attenuation coefficient, L is the communication distance, η_t indicates the transmission efficiency of the receiving antenna, and η_r indicates the transmit efficiency of the transmit antenna.

Figure 7.6 is a cross-sectional view of a Gaussian beam propagating in the atmosphere. Assuming that the beam travels in the z-axis direction, there is the following formula:

$$I(r, z) = I_0 \times \frac{W_0^2}{W_z^2} \times \exp\left[-\frac{2r^2}{W_z^2}\right] \tag{7.4}$$

In the formula, r is the distance from the vertical plane of the transmission path to point z, I_0 is the peak power at $z = 0$, W_0 is the spot radius of $z = 0$, called beam waist, and W_z is the spot radius at the z point.

Under normal circumstances, if the beam exhibits a Gaussian curve distribution, the expression of the change of the spot radius with the beam transmission distance is shown in formula (7.5):

$$W(z) = W_0\sqrt{1 + \left(\frac{\lambda z}{\pi W_0^2}\right)^2} \tag{7.5}$$

In the formula, λ represents the laser wavelength.

Due to the transmission characteristics of the beam, the beam formula (2.4) can be simplified as:

$$W(z) \approx \frac{\lambda z}{\pi W_0} \tag{7.6}$$

Thus, we get

$$\theta = \lim_{z \to \infty} \frac{W(z)}{z} = \frac{\lambda}{\pi W_0} \tag{7.7}$$

Placing θ and W_0 at the optical antenna, we get:

$$L_{geo} = \frac{P_r}{P_t} = \frac{\int_0^{d_r/2} I(z, r) \times 2\pi r dr}{P_t} = \frac{\frac{\pi I_0 W_0^2}{2} \times \left[1 - \exp\left(-\frac{d_r^2}{2W_z^2}\right)\right]}{P_t} \tag{7.8}$$

In the formula, d_r is the diameter of the receiver antenna, r is the distance from the vertical plane of the transmission path to point z, and W_0 is the spot radius at $z = 0$.

The outgoing power emitted by the light source after passing through the system transmitting antenna is shown in formula (2.8):

$$P_t = \int_0^{d_r/2} I(r, 0) \times 2\pi r dr = \frac{\pi I_0 W_0^2}{2} \times \left[1 - \exp\left(-\frac{d_t^2}{2W_0^2}\right)\right] \tag{7.9}$$

where d_t is the diameter of the transmitting antenna.

Due to atmospheric absorption, scattering, and other factors that will cause power attenuation, the following empirical formula can be obtained:

$$\alpha = \frac{3.912}{V}\left(\frac{0.55}{\lambda}\right)^q \tag{7.10}$$

V is the atmospheric visibility, λ is transmission laser wavelength, and q is the wavelength correction factor, which varies with atmospheric visibility, as shown in Table 7.2.

7.4.2 The Role of the Tracking System in the APT System

In the workflow of the APT system, the tracking link is after the capturing link is completed, and its function is to stabilize the captured target light spot at the receiving end to precisely track the specific range of the detector surface. Since the system is interfered by various factors, the position signal of the detected target spot is a jitter signal with a large residual error, and so it is necessary to reduce the residual error of the jittered target spot as much as possible through the tracking

Table 7.2 Correction factor corresponding to the different meteorology, where P_r must be greater than the sensitivity S of the receiver detector.

Correction factor	Visibility	Visibility level	Weather conditions
0	$V < 500$ m	<3	Dense fog
V–0.5	$0.5 < V| < 1$ (km)	3	Light fog
0.16 V + 0.34	$1 < V < 6$ (km)	4–6	Haze
1.3	$6 < V < 50$ (km)	6–8	Sunny
1.6	>50 (km)	9	Extremely sunny

stage, and also need to obtain in real time the coordinate parameters of the beam pointing to prepare for the subsequent alignment stage and the reception of communication signals [9, 10].

The alignment stage is the process of obtaining the arrival direction of the incident beam through the position deviation of the spot detected in the tracking stage, and then guiding the local laser beam to emit coaxially in the opposite direction of the detected beam direction, thus achieving the dual-mode APT system end alignment. The alignment and tracking phases of the beam are closely related and together determine the overall accuracy of the APT system [11–13].

Theoretically, it is assumed that the alignment error between the arrival direction of the beam and the optical antenna at the receiving end is $\pm\psi_a$; to perform stable and reliable detection at the receiving end detector, the beam width should be at least $\pm 2\psi_a$.

Assume the beam has a width of radians ψ_b, the power detected by the receiver detector is P_{rb}. To compensate for alignment errors in radians $\pm\psi_a$, increase the beam width to radians $\psi_b + 2\psi_a$, and the power detected by the receiver drops to:

$$P_t = P_{rb}\left(\frac{\psi_b}{\psi_b + 2\psi_a}\right) \tag{7.11}$$

It can be seen from formula (7.11) that a lot of power loss will occur at the receiving end.

The alignment relationship of wireless laser communication is shown in Figure 7.7.

As shown in Figure 7.7a, the light spot signal enters the coarse detection FOV through coarse alignment. As shown in Figure 7.7b, after completing the first-stage alignment work, the light spot signal enters the FOV for fine detection. It also indicates that the spot signal has entered the center of the detection FOV.

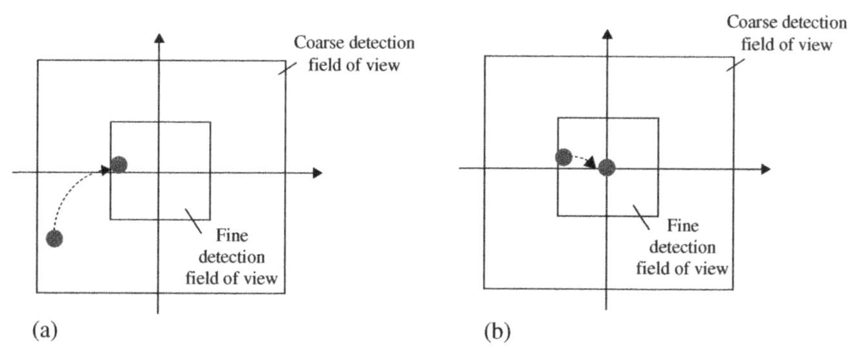

Figure 7.7 A diagram of the spot in the field of view. (a) Coarse alignment process; (b) fine alignment process.

In fact, when the system enters the fine alignment process, it also enters the tracking stage, that is to say, the tracking process is the cycle of the fine alignment process [12].

7.5 APT System Terminal Structure

For the inter-satellite laser communication system, it is also called APT system. In order to better realize the high-performance APT process, the APT system adopts the composite axis control structure of coarse and fine sighting as shown in Figure 7.8. Therefore, the APT system can be divided into coarse sighting subsystem and fine sighting subsystem. Among them, the task of the coarse aiming subsystem is to complete the acquisition, coarse aiming, and coarse tracking process of the inter-satellite laser communication, which is the guarantee for the inter-satellite laser communication system to realize the communication task. However, on the whole, the positioning accuracy after rough aiming is low, and the impact on the link stability in the later stage is also small. The precise aiming subsystem has a small movement range, which can achieve precise aiming and precise tracking of the communication target, high precision, and high bandwidth, which is the key to improving the tracking accuracy of the inter-satellite laser communication system and ensuring the stability of the link. Therefore, a composite axis control structure can be formed by combining the coarse sighting and fine sighting subsystems to complement each other's advantages.

Figure 7.9 shows the terminal structure diagram of the inter-satellite laser communication system. Among them, (a) is a periscope structure. The periscope terminal has a simple structure and is integrated, but its terminal antenna is only installed in the azimuth axis and the elevation axis. The two mirrors on the top and the three optical mirrors of the outer mirror on the pitch axis have weaker beam control accuracy than the frame type. The laser communication experimental terminal carried

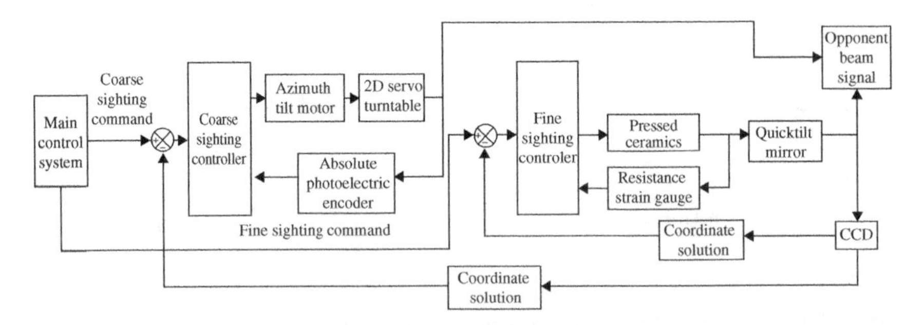

Figure 7.8 Schematic diagram of the composite axis control structure of the inter-satellite laser communication system.

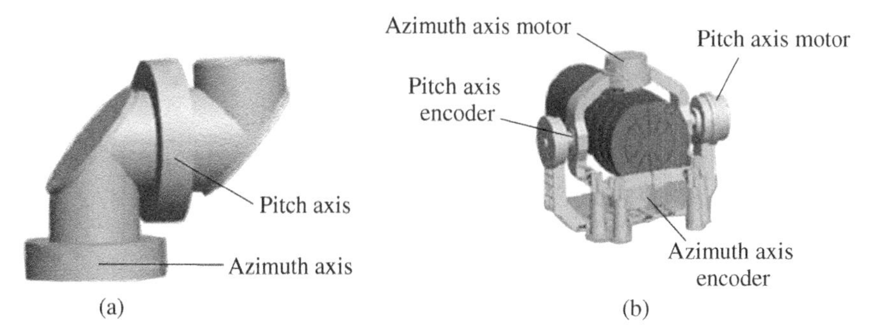

Figure 7.9 Terminal structure diagram of inter-satellite laser communication system. (a) Periscope structure; (b) frame structure.

on the "Ocean No. 2" satellite adopts a periscope structure because it belongs to a near-Earth exploration satellite. (b) It is a frame structure, the mechanical structure is more reliable, and the anti-interference ability is stronger, so it can realize more complex communication tasks. However, due to the internal and external frame structure, the requirements for the mechanical strength of the system materials are relatively high. The satellites carried by the terminal of the inter-satellite laser communication system in the second phase of the project are high-orbit satellites, which will achieve longer-distance communication transmission tasks, so the terminal adopts a frame structure.

7.5.1 Coarse Sight Subsystem Design

7.5.1.1 Coarse Sight Subsystem Composition

The main structure of the rough sight subsystem includes two parts: the main body of the inter-satellite laser communication terminal and the electronic control box, as shown in Figure 7.10. The terminal electronic control box includes the control body of the rough sight subsystem, and there are three units: Aim Up, Coarse Control, and Coarse Drive Units. The power required by each unit circuit board is provided by the power converter through the conversion chip in it.

The main body of the terminal includes actuators and sensors such as drive motors, two-dimensional servo turntables, photoelectric encoders, and coarse-sighted feedback units. The main workflow is shown in Figure 7.11.

1) **Photoelectric encoder**

   In the inter-satellite laser communication terminal system, the photoelectric encoder is used to measure the steering position information of the terminal antenna. There are light and dark coding patterns on the photoelectric coding disc, and the current steering position information can be obtained by reading the coding information through the reading head. According to the working

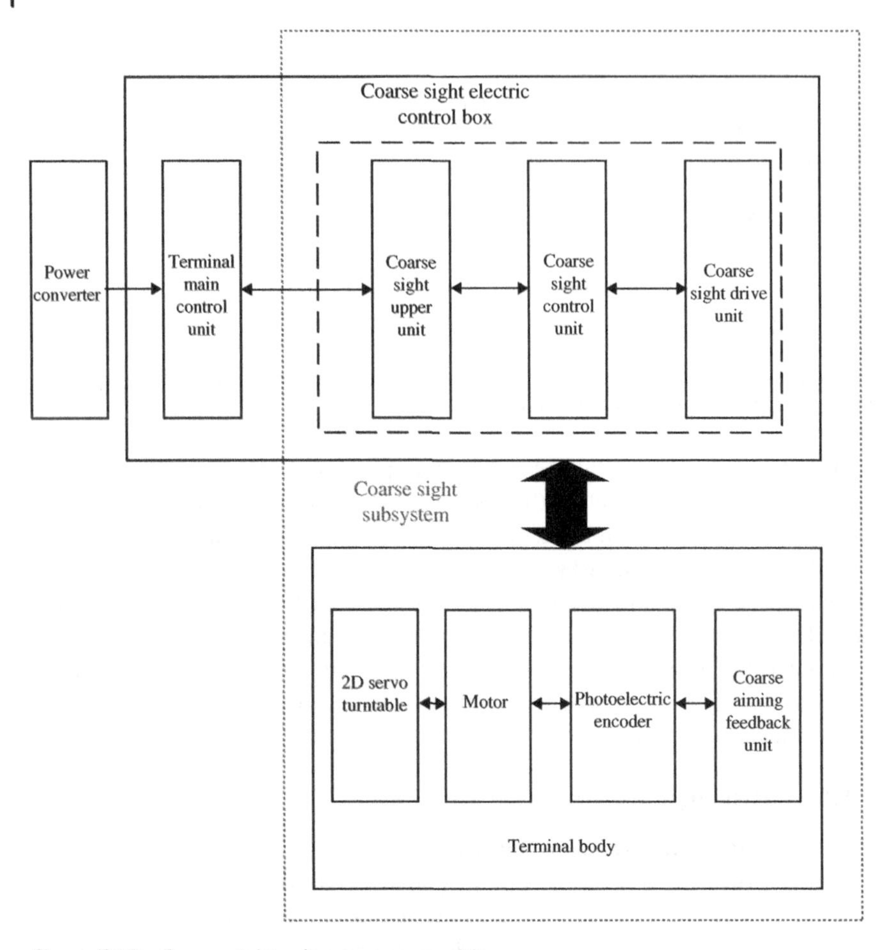

Figure 7.10 Coarse sight subsystem composition.

principle of the photoelectric encoder, the photoelectric encoder can be divided into two categories: absolute and incremental, as shown in Figure 7.12.

2) **Motor**

The magnetic field of the permanent magnet synchronous motor is provided by permanent magnets, so the motor structure is simple, reliable, and small in size. According to the application environment requirements of the inter-satellite laser communication system, the motor needs to have the characteristics of high efficiency, low power consumption, large starting torque, etc., and the permanent magnet synchronous motor just meets the requirements, so this system uses the permanent magnet synchronous motor to provide stable torque output control actuator.

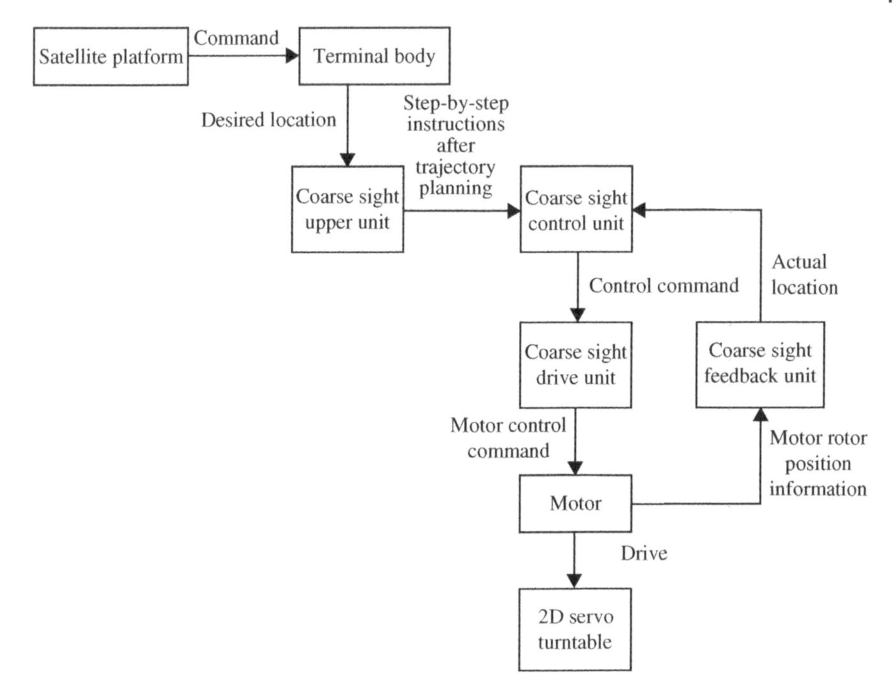

Figure 7.11 Coarse sight subsystem working flow chart.

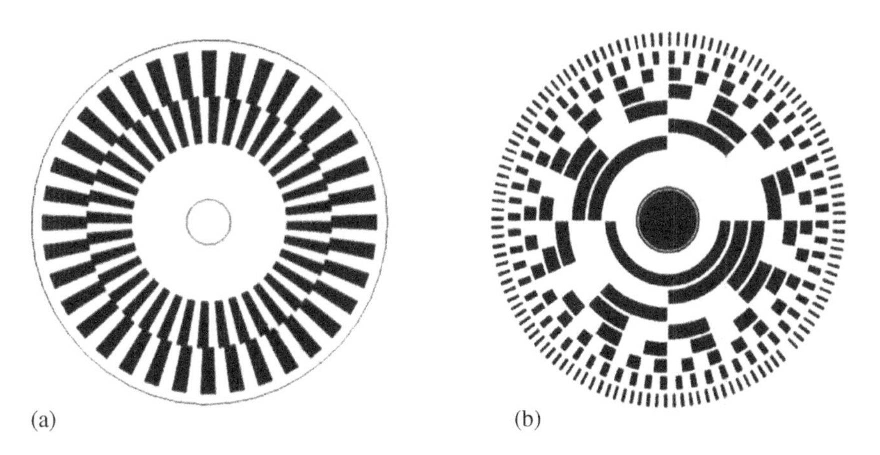

Figure 7.12 Incremental and absolute encoder markings. (a) Incremental; (b) absolute gray code.

3) **2D Servo turntable**

According to the previous analysis, in order to realize the communication task of high-orbit satellites, the incremental photoelectric encoder with internal and external frame-type mechanical structure needs to first select the reference position, and then provide the relative displacement based on the reference position. It also provides the absolute position of the motor rotor. The advantage of incremental photoelectric encoders is the principle structure is simple and easy to implement; the disadvantage is that there is a problem of data loss in power failure, and each time to obtain absolute position information, the previously measured data needs to be accumulated, so errors may be caused. The accumulation of information makes the deviation of the obtained displacement information larger. As for the absolute photoelectric encoder, as long as it is in the working state, the absolute position of the current motor rotor can be obtained according to the magnitude of the output photocurrent. Absolute photoelectric encoders have different coding methods, as shown in Figure 7.12b is an absolute photoelectric encoder with gray coding. According to the requirements of the project index, a 22-bit absolute photoelectric encoder is used.

7.5.1.2 Coarse Aiming Control Subsystem Design

As shown in Table 7.3, it is the control index requirements of the coarse sight subsystem.

According to the task of the coarse sighting subsystem, its working characteristics are large stroke and low bandwidth. Because the inner and outer frames are driven to rotate and the load is large, in order to reduce the moment of inertia and achieve more accurate alignment, a lower running speed is required. Therefore, the control task of the coarse sighting subsystem requires high-precision positioning and accurate speed tracking. According to the above performance indicators

Table 7.3 Coarse sight subsystem control indicators.

Parameter	Indicator requirements
Positioning error	≤3 μrad
Tracking error	≤30 μrad
Bandwidth	≥8 Hz
Standard acceleration	3.53 m/s^2
Peak acceleration	5.55 m/s^2
Tracking speed	≥0.5°/s
Scan speed	≥3.5°/s

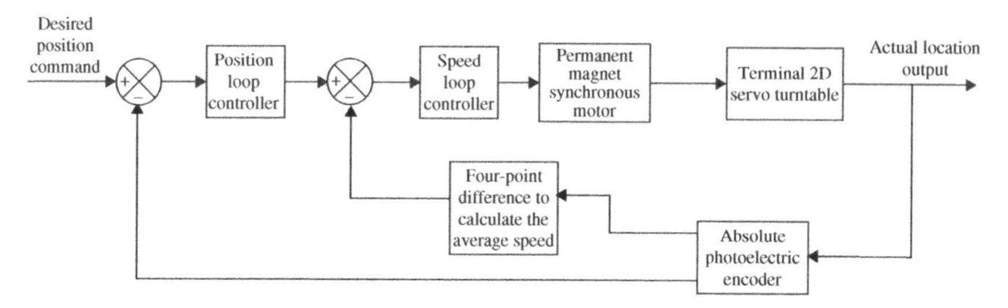

Figure 7.13 Coarse aiming control system block diagram.

and the whole control task, the block diagram of the coarse aiming control system is designed as shown in Figure 7.13, and the double closed loop control structure of position loop and speed loop is adopted.

7.5.2 Design of Precision Sighting Subsystem

7.5.2.1 The Composition of the Precision Aiming Subsystem

The task of the fine-sighting subsystem is to perform further fine-sighting and real-time stable fine-tracking on the target beam after the coarse-sighting subsystem completes the capture and coarse aiming of the target beam. Figure 7.14 shows the composition diagram of the precision sighting subsystem, which mainly includes the main optical system of the terminal and the precision sighting electronic control box. There are two units of precision sighting control and precision sighting drive in the precision sighting electronic control box. Because the driving of piezoelectric ceramics (PZT) requires a high voltage of ± 100 V, in order to avoid the influence on the control circuit, we will make the plate separately. The control voltage signal is amplified by the power of the driving board to obtain the driving voltage to drive the piezoelectric ceramic, and the resistance strain gauge is used to detect the displacement and deformation of the piezoelectric ceramic, and the displacement information is fed back to the control board. The optical system of the terminal body includes actuators and sensors such as PZT, fast tilt mirrors, and spot detectors, as well as optical equipment such as laser signal transmitters and receivers, beam splitters, primary mirrors, secondary mirrors, and lenses.

Each unit of the precision sight subsystem is provided with the required power by the power converter. Main work flow: when receiving the signal, control the PZT to drive the deflection of the fast tilting mirror to adjust the incident beam reflected by the primary mirror and the secondary mirror to the fast tilting mirror, and then pass through the beam splitter, and one part of it enters the information receiver and the other part of it enters the spot detection device; similarly, when the transmitter sends a signal, the piezoelectric ceramic is controlled to drive the

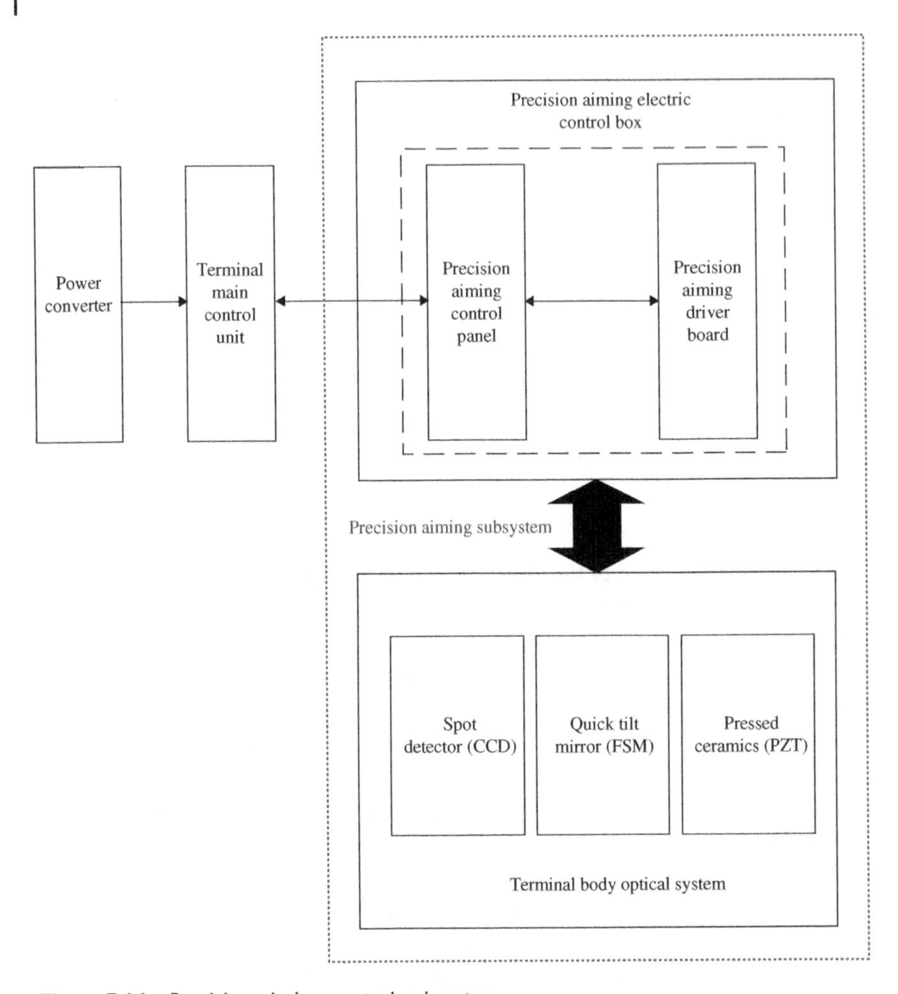

Figure 7.14 Precision aiming control subsystem.

fast tilt mirror to adjust the angle of the outgoing light, and then the beam is sent out through the reflection of the primary mirror and the secondary mirror.

1) **Fast Steering Mirror (FSM)**
   Fast Steering Mirror, FSM, is a key component of the precision sighting subsystem of the inter-satellite laser communication system, and it is the control object in the precision sighting subsystem. It adjusts the incident light and outgoing light through its own deflection to complete the precision sighting task. The fast tilt mirror has the advantages of small dynamic hysteresis error, fast response speed, and high resonance frequency, but the disadvantage is that the working stroke is small, which is in line with the control requirements

of the precision aiming system. There are two ways to connect the quick tilt mirror to the working platform: one is the flexible ring and hinge structure. The reflector is connected to the piezoelectric ceramic through the flexible hinge to realize the linkage between the piezoelectric ceramic and the reflector. The second is a fixed connection structure with the inner and outer frames. The mirror is fixed on the inner and outer frame structure, and the rotation of the azimuth axis and the pitch axis of the mirror is driven by the motor.

2) **Piezoelectric Ceramics (PZT)**

   There are generally two general drivers for fast tilt mirrors, one is PZT, whose principle is the inverse piezoelectric effect, and its displacement is controlled by the magnitude of the voltage applied to the PZT. Its characteristics are small size, high frequency response, large driving torque, and small stroke. The other is the voice coil motor, whose working principle is to control the magnitude and direction of the push–pull force by changing the magnitude and direction of the current of the coil set on the permanent magnet. It is characterized by a very good frequency response, large stroke, and so on. According to the control tasks and characteristics of the precision sighting subsystem, in this project, PZT are used as the driver of the fast tilting mirror.

   PZT have hysteresis characteristics and creep characteristics in the working process, because this is determined by the internal material, and so this is the inherent characteristics of PZT. As a result, the control difficulty of PZT is increased. Among them, the voltage–displacement error caused by hysteresis in the control process can reach 15–20%, and the error caused by creep characteristics is 1–5%.

3) **Piezoelectric deflection mirror system**

   PZT connect the mirror through flexible hinges, and the mechanism that drives the mirror to deflect is called a piezoelectric deflection mirror system.

4) **Charged Coupled Device (CCD)**

   Charged Coupled Device, CCD, is a semiconductor device that converts optical signals into electrical signals and obtains optical signal data by collecting electrical signals. The main working process is that the irradiation of the communication beam causes the electrons to obtain energy transition, generate charges, collect, store, and transmit the charges, and finally detect the charge data to obtain the irradiation beam information. The intensity of light is characterized by the number of charges, and finally the spot image data is obtained by detecting the number of charges. CCD is generally divided into linear CCD and area CCD.

7.5.2.2 Design of Precision Aiming Control System

To sum up, the precise sighting subsystem needs to control the deflection of the PZT to drive the fast tilting mirror to adjust the beam angle, so it is necessary to control the displacement of the PZT and the beam deflection angle, and the

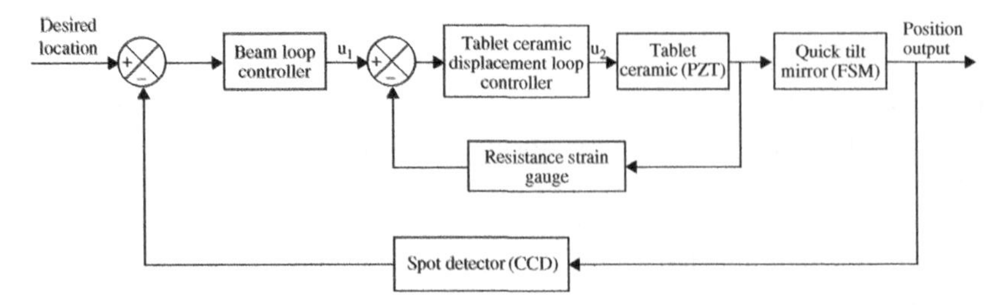

Figure 7.15 Double-loop control system block diagram of precision aiming.

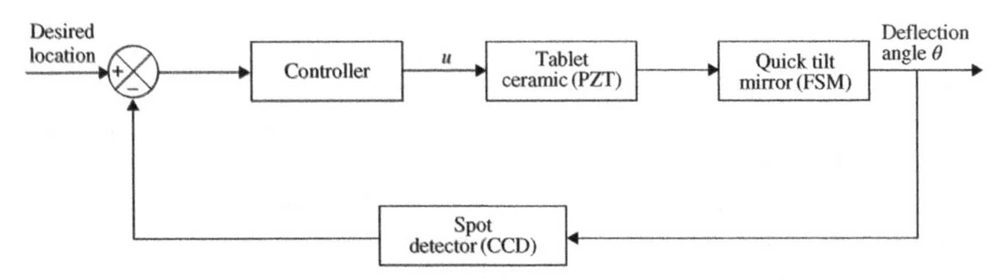

Figure 7.16 Precision aiming control system block diagram.

existence of hysteresis and creep characteristics makes it possible to precisely control the piezoelectricity. Ceramic displacement becomes complex and requires a separate displacement control loop. But our ultimate goal is to control beam deflection, so an outer loop of beam feedback is needed. Therefore, the entire precision aiming control system adopts a double-loop closed-loop control structure, including an inner loop of piezoelectric ceramic displacement control and an outer loop of beam control, as shown in Figure 7.15. This paper will mainly discuss the performance analysis of the precision sighting system under external interference, so the hysteresis and creep characteristics of the PZT inside the precision sighting subsystem will not be considered, that is, without considering the PZT displacement control loop, Figure 7.16 shows the block diagram of the precise sighting control system.

References

1 Knibbe, T.E. (2016). Spatial tracking using an electro-optic nutator and a single-mode optical Fiber. *Proceedings of SPIE* 35 (16): 309–317.

2 Komukai and Toshibacho (2014). Performance evaluation of laser communication equipment. *Proceedings of SPIE* 36 (92): 41–50.

3 Roberts, W.T. and Wright, M.W. (2013). The lunar laser OCTL terminal (LLOT) optical systems. *Proceedings of SPIE* 32 (43): 11–19.
4 Tolker-Nielsen, T. and Oppenhaeuser, G. (2015). In-orbit test result of an operational optical inter satellite link between ARTEMIS and SPOT4, SILEX. *Proceedings of SPIE* 46 (35): 1–15.
5 Sodnik, Z., Lutz, H., and Furch, B. (2016). Optical satellite communications in Europe. *Proceedings of SPIE The International Society for Optical Engineering* 75 (87): 87–97.
6 Pease, R. (2015). Optical laser communication system scar venichein metromarkets. *Light Wave* 45 (9): 23–27.
7 Komukai, S. (2014). Performance evaluation of laser communication equipment onboard the ETS_VI satellite. *Proceedings of SPIE* 36 (92): 41–50.
8 Arnon, S. (2016). Optimum transmitter optics aperture for free space satellite optical communication. *Proceedings of SPIE* 28 (11): 252–264.
9 Kenichi, Janhaiku, and Saiwaiku (2017). Experimental operations of laser communication equipment on board ETS_VIsatellite. *Proceedings of SPIE* 29 (90): 264–275.
10 Francois (2014). Simulation model and on-ground performancesvalidation ofthe PAT system for SILEX program. *Proceedings of SPIE* 14 (17): 262–276.
11 Borrello, M. (2005). Multi stage pointing acquisition and tracking control system approach. *Proceedings of the American Control Conference* 23 (6): 75–80.
12 Gallagher, N. Jr. (1991). A theoretical analysis of the median filters. *IEEE Transactions on Acoustics, Speech, and Signal Processing* 29 (1): 1136–1141.
13 Ballard, D.H. (1981). Generalizing the hough transform to detect arbitrary shapes. *Pattern Recognition* 1 (s2): 111–122.

8

Inter-Satellite Laser Link Tracking Error

8.1 Definition of Alignment Error

Positioning and tracking in the ATP system is one of the most important links, which has attracted more and more attention of scholars at home and abroad. The alignment error, as a measure of the effectiveness of various positioning algorithms, has always been the key to the analysis and research of scholars. The essence of the fine tracking process of the ATP system is the tracking and alignment process of the beam. In the ATP system, adjusting the boresight of the receiving end so that the received laser beam can accurately enter the designated position is the alignment process. After the alignment process is completed, the angular deviation between the aligned position and the actual center position relative to the transmitting end is the alignment error, which is a physical quantity to measure the accuracy of the ATP system. If the error of this quantity is large, the ATP system's accuracy must be lower. It can be seen from the above definition that the alignment error is essentially the deviation of the actual boresight from the target boresight. In the actual alignment process, due to the influence of factors such as platform motion and vibration, external interference, internal noise of the system, and the characteristics of the beam itself, the error value is always not equal to 0, and has a certain randomness. Therefore, the general alignment error is generally determined by a random variable; it needs to be studied by means of mathematical statistics [1].

According to the location of the alignment error, it can be divided into the transmitter aiming error and the receiver tracking error. The transmitter aiming error includes the advance aiming error and calibration error, and the receiver tracking error includes the controller residual error and alignment information acquisition error. Generally, the antennas of the inter-satellite ATP system are shared systems for transceivers. If the forward alignment of the transmitter is not considered, the alignment error of the transmitter and the alignment error of the receiver are

Laser Inter-Satellite Links Technology, First Edition. Jianjun Zhang and Jing Li.

Published 2023 by John Wiley & Sons, Inc.

equivalent. If the forward alignment of the transmitter is considered, the forward alignment angle is determined by the propagation speed of the laser and the lateral motion between satellites, so the forward alignment error is a static bias error, while other transmitter alignment errors are related to the receiver. The end alignment errors are still partially comparable and should all be considered as random dynamic errors. The following work in this chapter is to ignore the influence of the leading alignment error and analyze the mathematical model of the random dynamic error [2, 3].

8.2 Alignment Error Model and Factor Analysis

8.2.1 Mathematical Modeling of Alignment Errors

Figure 8.1 is a schematic diagram of the alignment error of laser long-distance transmission. Assuming that Ω is a point in the circular transmission antenna, the coordinates of this point are defined as (x_1, y_1), then the light intensity of this point can be:

$$I_t(t, x_1, y_1) = \begin{cases} I_1(t_1); & x_1^2 + y_1^2 \leq (d_t/2)^2 \\ 0; & elsewhere \end{cases} \tag{8.1}$$

Here, $t_1 = t + (z/c)$ is defined as the transmission delay of the laser signal, d_t is the transmission aperture, and c is the propagation speed of the light. Here, the emitted light intensity at time t can be expressed as:

$$I_1(t_1) = \left(\frac{I_1}{\sqrt{A_t}}\right) m(t) \exp\left[j(\omega t + \phi_1(t))\right] \tag{8.2}$$

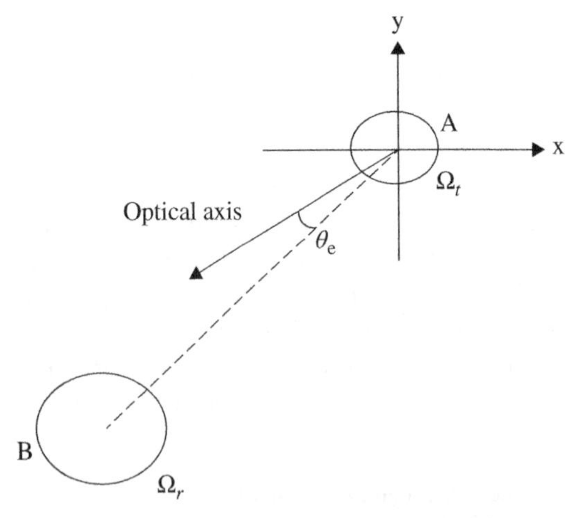

Figure 8.1 Schematic diagram of alignment error.

where A_t is the area of the transmitting antenna, $\left(I_1/\sqrt{A_t}\right)$ is the normalized field amplitude, $m(t)$ is the modulation wavelength ($|m(t)| \leq 1$), ω is the angular frequency of the optical carrier, $\phi_1(t)$ is the random distortion phase associated with the optical device, h is Planck's constant, and $v = \omega/2\pi$ is the optical carrier frequency, which $\left(I_1/\sqrt{A_t}\right)$ can be interpreted as the average photon velocity produced by an equivalent field amplitude $n_t = I_1^2/hv$ in the transmitting antenna area, the angle between the straight line formed by the center of the transmitting and receiving antennas in Figure 8.1, and the optical axis emitted by the transmitting end θ_e is the alignment error.

In Figure 8.1, A is the transmitter and B is the receiver. The light intensity distribution of the laser emitted from the A end to the B end after long-distance transmission should satisfy the Fraunhofer diffraction. Considering that the spot coverage area is much larger than the antenna size, the phase difference and spot amplitude on the receiving antenna are negligible. The light intensity distribution at end B can be expressed as:

$$I(t, R, \theta) = \beta I_A(t) f^2(z) G^2(\theta_e) \tag{8.3}$$

where β is the antenna gain and loss at the transceiver end, $I_A(t)$ is the transmitted light intensity at time, and $f(z)$ is the distance loss:

$$f(z) = \frac{e^{j2\pi z/\lambda}}{j\lambda z} \tag{8.4}$$

where λ is the laser wavelength and z is the distance between the center of the transmitting and receiving antennas.

In formula (8.4), $G(\theta)$ is the normalized diffraction amplitude of the transmitting antenna, which can be expressed as:

$$G(\theta_e) \approx \frac{2J_1(\pi d_t \theta/\lambda)}{\pi d_t \theta/\lambda} \tag{8.5}$$

where $J_1(\cdot)$ is the first-order Bessel function, d_t is the transmit antenna aperture, and θ_e is the boresight error angle.

Orthogonally decompose the boresight error θ_e angle into θ_x and θ_y:

$$\theta_x = \theta_e \cos\psi,\ \theta_y = \theta_e \sin\psi,\ \psi = \tan^{-1}\left(\frac{\theta_y}{\theta_x}\right) \tag{8.6}$$

Suppose θ_x and θ_y are independent random variables whose mean values are φ_x, φ_y, and variances are σ_x^2, σ_y^2, respectively, then the alignment error can be expressed as:

$$\theta(t) = \left[\theta_x^2(t) + \theta_y^2(t)\right]^{½},\ \varphi(t) = \left[\varphi_x^2(t) + \varphi_y^2(t)\right]^{½} \tag{8.7}$$

When analyzing the alignment error in azimuth and pitch in space optical communication, it can be assumed to satisfy the Gaussian distribution. Set $\sigma_x = \sigma_y = \sigma$, then the joint probability density function of θ_x and θ_y is:

$$p(\theta_x, \theta_y) = p(\theta_x) \cdot p(\theta_y) = \frac{1}{\sqrt{2\pi\sigma^2}} e \frac{\theta_x^2 + \theta_y^2}{2\sigma^2} \tag{8.8}$$

Then, the joint probability distribution of random variables θ_x and θ_y is:

$$P(\theta_x \leq \theta_X, \theta_y \leq \theta_Y) = \int_{-\infty}^{\theta_x} \int_{-\infty}^{\theta_Y} p(\theta_x, \theta_y) d\theta_x d\theta_y \tag{8.9}$$

Starting from the above basic mathematical model, combined with the dynamic random nature of the alignment error of space optical communication, the horizontal, pitch, and radial alignment errors are deduced from a statistical point of view.

Since the double integral from plane coordinates to polar coordinates has the following conversion relationship:

$$\iint_D f(x,y)dxdy = \iint_D f(x(u,v), y(u,v)) \left| \frac{\partial(x,y)}{\partial(u,v)} \right| dudv \tag{8.10}$$

where $|\partial(x, y)/\partial(u, v)|$ is the Jacobian determinant, and the above formula can be written as

$$\iint_D p(\theta_x, \theta_y) d\theta_x d\theta_y = \iint_D p(\theta_x(\theta, \varphi), \theta_y(\theta, \varphi)) \left| \frac{\partial(\theta_x, \theta_y)}{\partial(\theta, \varphi)} \right| d\theta d\varphi \tag{8.11}$$

where,

$$\left| \frac{\partial(\theta_x, \theta_y)}{\partial(\theta, \varphi)} \right| = \begin{vmatrix} \cos\varphi & -\theta\sin\varphi \\ \sin\varphi & \theta\cos\varphi \end{vmatrix} = \theta$$

We can get: $P(\varepsilon < \theta_e) = \int_0^\theta \frac{\varepsilon}{\sigma^2} e \frac{\varepsilon^2}{2\sigma^2} d\theta_e$.

Therefore, the probability density function of the radial alignment error random variable is:

$$p(\varepsilon) = \frac{\varepsilon}{\sigma^2} e \frac{\varepsilon^2}{\sigma^2} \tag{8.12}$$

Integrate Eq. (8.12) as follows:

$$P(\varepsilon > \theta^0) = \int_{\theta^0}^{\infty} \frac{\varepsilon}{\sigma^2} e \frac{\varepsilon^2}{2\sigma^2} d\varepsilon \tag{8.13}$$

where θ^0 is the instantaneous radial alignment error. Eq. (8.12) shows that the error probability that the instantaneous alignment error is greater than a certain alignment error ($\varepsilon > \theta^0$) threshold obeys the Rayleigh distribution [4].

8.2.2 Factors Causing Alignment Errors

After the capture is successful, the receiving end calculates the spot image position according to a certain positioning and tracking algorithm, takes this position as the position of the beacon light, compares it with a certain standard position, obtains the alignment information, and then uses the controller to adjust the receiving end's position. Look at the boresight and let the other party's beacon light reach the center of your own detector. In this process, due to the interference of its own optical equipment and the external environment, a certain alignment error occurs in the positioning accuracy. The following briefly introduces the error sources that affect the alignment accuracy in the inter-satellite ATP system.

As shown in Figure 8.2, the main sources of alignment errors include the degree of image preprocessing, the limitations of positioning methods, the influence of antenna aberrations on alignment, the relative motion between satellites, and the ability of the controller to control the servo mechanism, accuracy, vibration of the satellite platform itself, etc. [5].

In the inter-satellite ATP system, image preprocessing is to reasonably select the target region of interest (ROI) while removing noise interference. The purpose of denoising processing is to improve the signal-to-noise ratio of the image; the essence of the ROI window setting is to reduce the size of the laser spot image, removes unnecessary noise areas, and it also improves the signal-to-noise ratio of the image in disguise. According to previous research, the long-distance laser

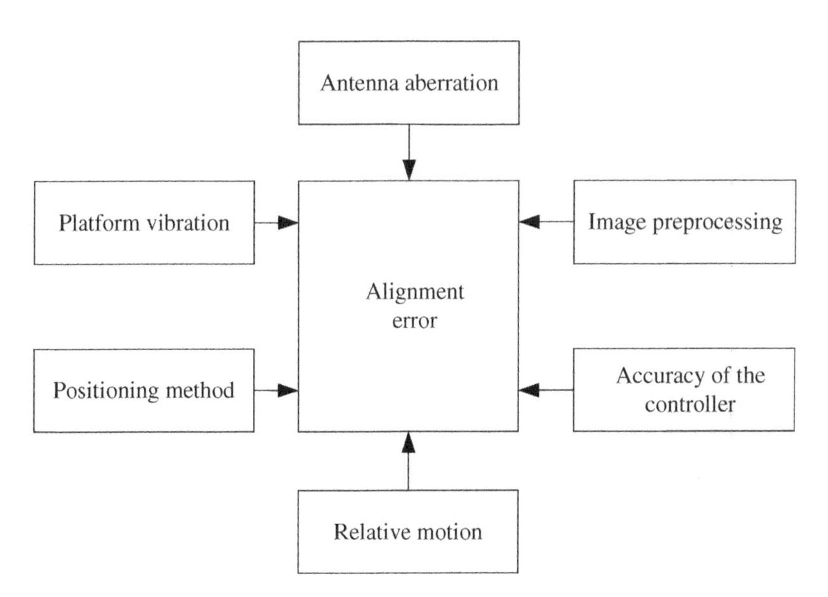

Figure 8.2 The main components of factors that cause alignment errors.

spot image is not only seriously disturbed by noise, but also occupies a small proportion of the entire CCD image. Therefore, laser spot image preprocessing is very necessary to suppress alignment errors and improve positioning accuracy. The positioning method is the soul of tracking. It goes without saying that the positioning accuracy of the algorithm itself must be the main source of the alignment error. Therefore, improving the positioning algorithm to improve the alignment accuracy is a topic worthy of study. Antenna aberration will lead to the distortion of the long-distance transmission laser, and then form a distorted spot and cause alignment error, so the analysis and suppression of the aberration is also the key to the ATP system [6].

8.2.3 Influence of Tracking Error on Beam Distribution at Receiver

The tracking error, that is, the difference between the current pointing angle of the terminal antenna of the inter-satellite laser communication system and the pointing angle of the target, as shown in Figure 8.3, namely

$$\Delta\theta = \overrightarrow{d_{SJ}} - \overrightarrow{d_{QW}} \tag{8.14}$$

Due to the inaccuracy of the control algorithm, the vibration of the satellite platform, the external and internal noise, and the performance of the terminal device, the abovementioned error value is not zero, which is the reason for the existence of the tracking error. In order to better study the effect of tracking error on the performance of the inter-satellite laser communication system, it is necessary to provide specific evaluation indicators. This paper proposes that it can be measured by the beam distribution at the receiving end and the communication bit error rate (BER). Therefore, the following two aspects will be analyzed in detail [7, 8].

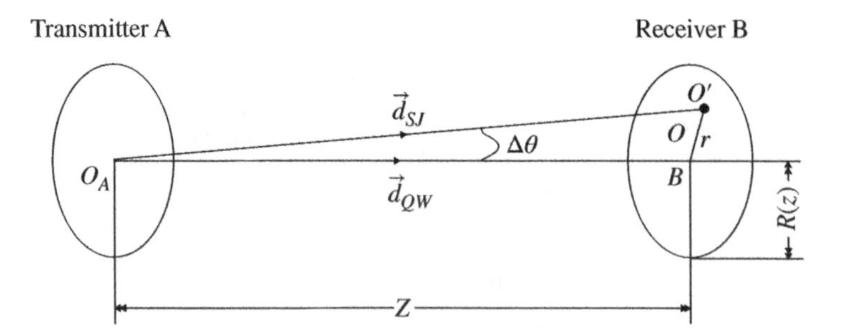

Figure 8.3 Schematic diagram of tracking and aiming error.

8.2.3.1 The Effect of Tracking Error on the Beam Intensity at the Receiving End

The inter-satellite laser communication system generally uses a Gaussian distributed laser as the communication beam. As shown in Figure 8.3, assume that the distance from the antenna center of transmitting terminal A (O_A point) to the antenna center of receiving terminal B (O_B point) is Z, and form a beam area with a radius of $R(z)$ at receiving terminal B; The light intensity at the center of the area (O_B point) is I_0, I_0' is the light intensity of O point on the area surface, r is the distance from O′ point to central O point.

For Gaussian beams:

$$I_{O'} = I_O e^{\left[-\frac{2r}{R^2(z)}\right]} \tag{8.15}$$

where:

$$R(z) = R_0\left[1 + \left(\frac{\lambda Z}{\pi R_0^2}\right)^2\right]^{\frac{1}{2}} \tag{8.16}$$

Among them, R_0 is the waist spot radius of the communication beam, and λ is its wavelength.

Assume that in the process of communication, the target pointing of terminal system A is $\overrightarrow{d_{QW}}$, but due to the existence of various interference factors, both parties in the communication have a pointing deviation, that is, the current pointing of terminal system A is $\overrightarrow{d_{SJ}}$, resulting in a tracking error $\Delta\theta$. The center of the beam at the receiving terminal antenna is located at point O', because the distance between point O' and point O is much smaller than r compared to the distance Z between the two communicating parties, so the tracking error at this time can be obtained as:

$$\Delta\theta \approx \frac{r}{z}$$

Then the light intensity at the center of the antenna of the receiving terminal B is:

$$I'_O \approx I_0 e^{\left[-\frac{2r}{R^2(z)}\right]} \tag{8.17}$$

In the same application scenario, except r, all are unchanged, so it can be normalized to a constant, and then the relation curve between the relative light intensity at the center of the receiving terminal and the tracking error angle can be obtained as shown in Figure 8.4.

As can be seen from Figure 8.4, the tracking error has a huge impact on the light intensity received at the center of the antenna at the receiving end. As the tracking

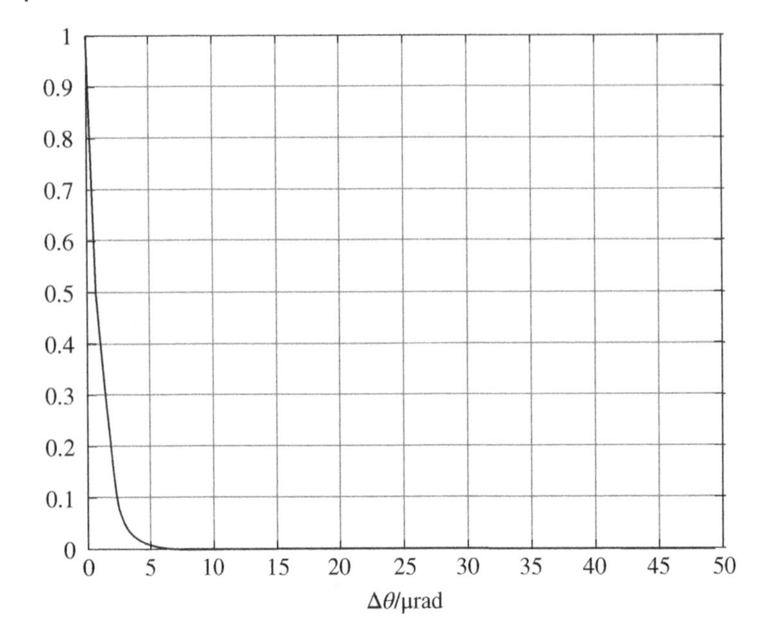

Figure 8.4 Relationship between relative light intensity and tracking error.

error increases, the light intensity at the center of the receiving end will decrease sharply, not only for determination of the center position of the beacon light spot and the reception of the communication signal light are both very unfavorable effects [9, 10].

8.2.3.2 Influence of Tracking Error on Beam Power at Receiver

According to the previous section, in the inter-satellite laser communication system, the laser with Gaussian distribution is used as the communication light source. Next, the influence of the tracking error on the beam power of the receiving end will be analyzed. Assume that the tracking error of the sending terminal is ψ, the receiving terminal will The tracking error is ϕ, then there are:

$$P = \frac{4aS\cos\phi}{r^2\alpha^2} e^{\left(-\frac{8\psi^2}{\alpha^2}\right)} \tag{8.18}$$

Among them, P is the optical power of the signal received by the receiving terminal, a is a constant, which is related to the laser properties, S is the antenna area of the receiving terminal, α is the beam divergence angle, and r is the distance between the transmitting terminal and the receiving terminal. In order to measure the influence of the tracking error of the transmitter and the receiver on the optical power of the signal received by the receiver, two quantities are defined: (ψ) and

$L(\phi)$, which represent the power influence factors of the transmitter and the receiver, respectively; from formula (8.18), it can be known that:

$$\begin{cases} L(\psi) = e^{\left(-\frac{s\varphi^2}{\alpha^2}\right)} \\ L(\phi) = \cos\phi \end{cases} \tag{8.19}$$

Figure 8.5 shows the influence of tracking and aiming error on the optical power of the receiver signal, and $L(\psi)$, $L(\phi)$ are normalized values. The smaller the value, the greater the adverse effect. As can be seen from the figure, as the tracking error increases, $L(\phi)$ is basically unchanged, indicating that the tracking error at the receiving end has little effect on the signal optical power received by the receiving end. However, with the increase of the tracking error, $L(\psi)$ decreases sharply, indicating that the tracking error of the transmitter has a great influence on the signal optical power of the receiver. This is an important evaluation reference information for the design of the inter-satellite laser communication terminal control system [11, 12].

8.2.4 Influence of Tracking and Pointing Error on Communication Error Rate

The influence of tracking error on the beam distribution received by the receiving terminal is studied above. The biggest influence is the feedback information of the beam pointing angle in the control process of the precision sighting system. Next, we will analyze the impact of tracking error on the BER. The BER is the main parameter to measure the system performance, and it will affect the communication performance of the system to a great extent [13].

In order to facilitate the research, this paper will choose the binary unipolar code as the communication information code. The principle of the binary unipolar code is: when there is no tracking error, the light intensity of the beam received at the center of the receiving terminal when the transmitting terminal sends a "1" code is the resulting beam intensity is zero. If there is a tracking error, we know from the previous section that the tracking error will affect the beam distribution received by the receiving terminal, thereby reducing the beam intensity. Then it must appear that the light intensity of the beam received at the center of the receiving terminal when the "1" code is transmitted is less than. We will use $I^* = (I_0/2)$ as the light intensity judgment threshold. When receiving light intensity $I < I^*$, it will be misjudged as "0" code.

Due to the tracking and aiming error, when the light intensity of the beam received at the center of the receiving terminal becomes I^*, the tracking and aiming error is $\Delta\theta = \theta^*$. From Eqs. (8.15)–(8.17), we can get:

$$\theta^* = \frac{R(z)\sqrt{\ln 2}}{\sqrt{2}z} \tag{8.20}$$

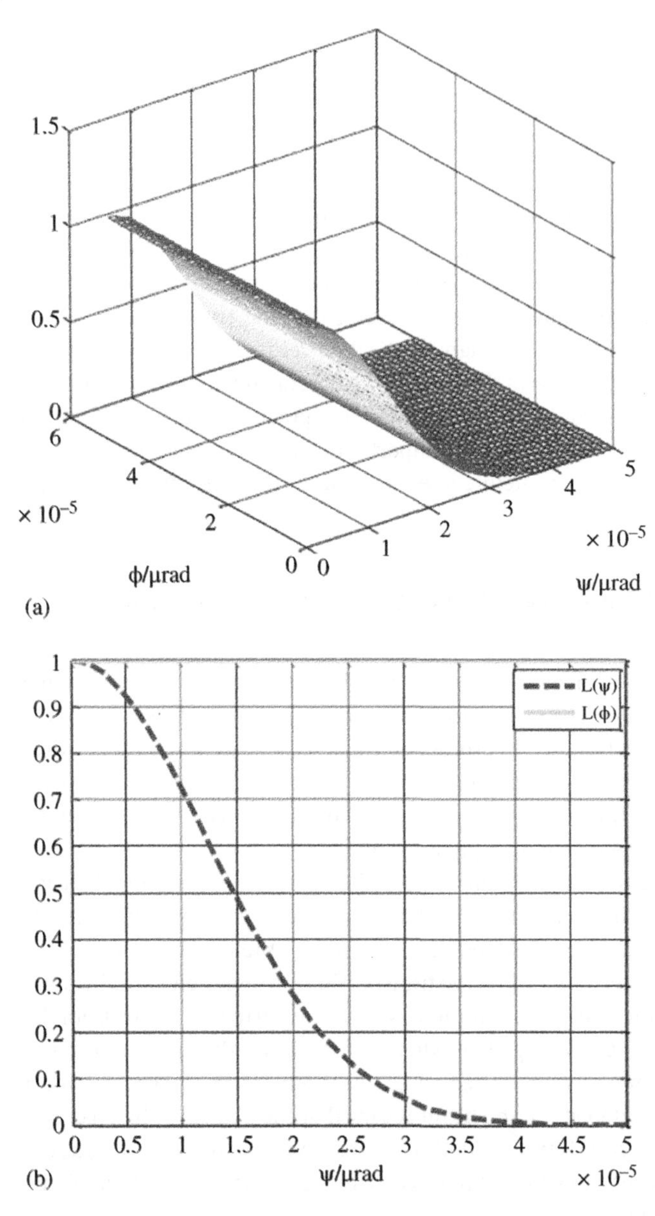

Figure 8.5 The effect of tracking error on the signal optical power at the receiving end. (a) Aiming error affects 3D map; (b) influence factor.

Because the communication beam of the inter-satellite laser communication system generally adopts a Gaussian distribution beam I^*, the probability density distribution function of the tracking error of the system can be obtained:

$$p(\Delta\theta) = \frac{1}{\sqrt{2\pi}\sigma} e^{-\frac{(\Delta\theta)^2}{2\sigma^2}} \tag{8.21}$$

where σ is the standard deviation of the tracking error.

The communication BER that the transmitting terminal sends a "1" code and the receiving terminal misjudges it as a "0" code can be obtained from the following formula:

$$\begin{aligned} BER &= \frac{1}{2}BER(0|1) + \frac{1}{2}BER(1|0) = \frac{1}{2}BER(1|0) \\ &= \int_0^{\infty} p(\Delta\theta)d(\Delta\theta) \\ &= \int_0^{\infty} p(\Delta\theta)d(\Delta\theta) - \int_0^{\theta^*} p(\Delta\theta)d(\Delta\theta) \\ &= \frac{1}{2}\left(1 - erf\left(\frac{\theta^*}{\sqrt{2}\sigma}\right)\right) \end{aligned} \tag{8.22}$$

where $erf(\cdot)$ is the error function. From formulas (8.20 and 8.21), we can get

$$BER = \frac{1}{2}\left\{1 - erf\left[\frac{R_0\left[1 + \left(\lambda z/\pi R_0^2\right)^2\right]^{\frac{1}{2}}\sqrt{\ln 2}}{2\sigma z}\right]\right\} \tag{8.23a}$$

It can be seen from formula (8.23a) that when other system parameters remain unchanged, the communication BER is only related to the standard deviation of the tracking error, thus the curve shown in Figure 8.6 can be obtained.

It can be seen from Figure 8.6 that the BER has a certain tolerance for the tracking error, but after a certain increase, it will rise sharply with the increase of the tracking error, which will seriously affect the communication quality.

8.3 Analysis of Tracking and Pointing Error Sources of Inter-Satellite Laser Communication System

The definition of the tracking and aiming error of the inter-satellite laser communication system was introduced earlier, and the impact of the tracking and aiming error on the system performance was analyzed from two aspects. It can be seen from this that the existence of the tracking and aiming error has an adverse effect on the system performance. It may even disrupt communication. Therefore, when designing an inter-satellite laser communication system, the tracking error must be suppressed or eliminated, but there are many reasons for the tracking error, and

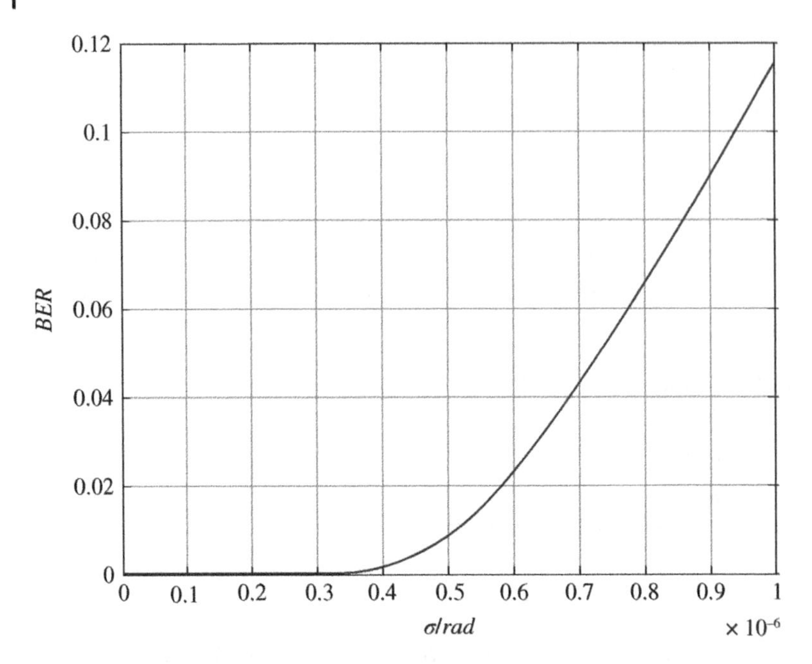

Figure 8.6 The relationship between tracking error and bit error rate.

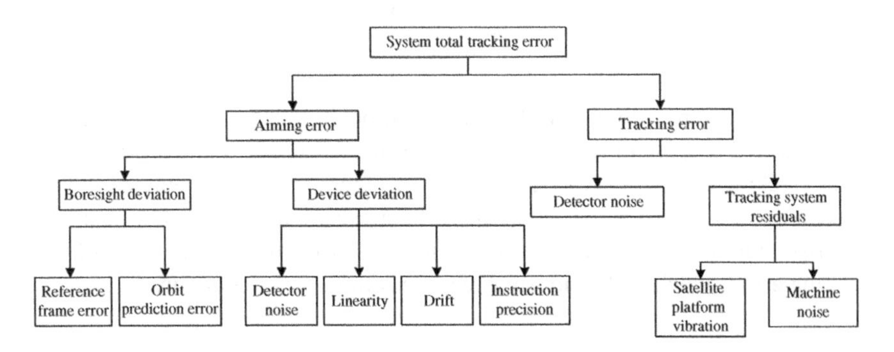

Figure 8.7 Tracking error source of inter-satellite laser communication system.

the types are different. Therefore, in order to better suppress the tracking error, we must first analyze the cause of the tracking error in detail, that is, the source of the error as shown in Figure 8.7.

It can be seen from Figure 8.7 that the total tracking and aiming error of the inter-satellite laser communication system is divided into aiming error and tracking error. Aiming error can be divided into boresight deviation and device error according to type. The factors that cause device error are detector noise, linearity, device parameter temperature drift and command accuracy, etc. This book does not discuss the performance of devices involved in linearity, device parameter temperature drift, instruction accuracy, etc., while LOS deviation is an avoidable algorithm error problem, which is not covered in this book. Tracking errors include detector noise, mechanical noise, satellite platform vibration, etc. For mechanical noise, it involves mechanical structure, and this paper will not discuss it. JPL uses the ATLAS simulation tool to assign indicators to the various sources of tracking errors in the inter-satellite laser communication system, the largest of which are satellite platform vibration, sensor random noise, and detector noise. Combining various error sources, satellite platform vibration and detector noise are the main factors that affect the tracking and aiming accuracy. Therefore, this chapter will analyze and study the satellite platform vibration and detector noise of the inter-satellite laser communication system, respectively, and propose suppression or compensation algorithms to improve the performance of the inter-satellite laser communication system and improve the system design. Research is significant [8, 9].

8.3.1 Satellite Platform Vibration

From the above analysis, it can be seen that the vibration of the satellite platform is the most influential source of the tracking error. Therefore, the analysis and research on the vibration of the satellite platform is the key to solving the tracking error of the inter-satellite laser communication system. The following will analyze the sources and characteristics in detail.

There are two main sources of vibration of satellite platforms, as shown in Table 8.1.

Among the vibrations generated by the interference of the external space environment, the impact of micrometeorites is an accidental event, so it is not considered, and the weak deformation of the rigid body can be classified as the mechanical motion of the satellite platform itself. Therefore, this paper believes that among the factors affecting the vibration of satellite platforms, internal reasons are dominant [3, 11].

The Japan National Aeronautics and Space Administration (NASDA) conducted actual measurement on the satellite ETS-VI, and obtained the measured data of satellite vibration, as shown in Table 8.2.

NASA obtained the vibration power spectral density of the satellite platform by measuring the LANDSAT satellite, as shown in Figure 8.8.

From the above analysis of the vibration of the satellite platform, it can be concluded that the vibration of the satellite platform is characterized by high amplitude at low frequency and low amplitude at high frequency, so the impact on the performance of the inter-satellite laser communication system is mainly the vibration of the low frequency part. To be precise, it is the part below 100 Hz.

Table 8.1 Vibration types and their effects on the inter-satellite laser communication system.

Source	Influencing factors	Feature
The satellite platform operates mechanically on its own	Propeller vibration noise Reaction wheel operating noise Thrust ring operating noise Antenna mechanical movement Fast mirror rotation in the control system Solar array drive noise	It is concentrated in the middle and low frequency domain, and it is more prominent at individual frequency points. The main influencing factors are the operating noise of the propeller and the mechanical movement of the antenna.
Disturbed by the external space environment	Impact of micrometeorites Gravitational noise of space stars such as the Earth and the moon Weak deformation of rigid body caused by satellite in space environment temperature change Rigid body weak deformation Solar radiation pressure	Random vibration with small amplitude but wide spectrum. The main influencing factors are the impact of micrometeorites and the weak deformation of rigid bodies caused by satellites under the change of space environment temperature

Table 8.2 Satellite vibration angles at different sampling frequencies.

Sampling frequency/Hz	Pitch axis/μrad	Scroll axis/μrad	Radial/μrad
500 Hz	7.3	5.13	8.9
100 Hz	11.2	18.6	21.7
1 Hz	The maximum vibration amplitude is 200 μrad		

8.3.2 Detector Noise

8.3.2.1 Characteristics and Types of Detector Noise

From the analysis of the design of the entire inter-satellite laser communication system in Chapter 2, we can see that in the fine-sighting subsystem, we use the area array CCD as the feedback sensor of the fine-sighting control system, that is, the spot detector, which detects the spot position information and feeds back

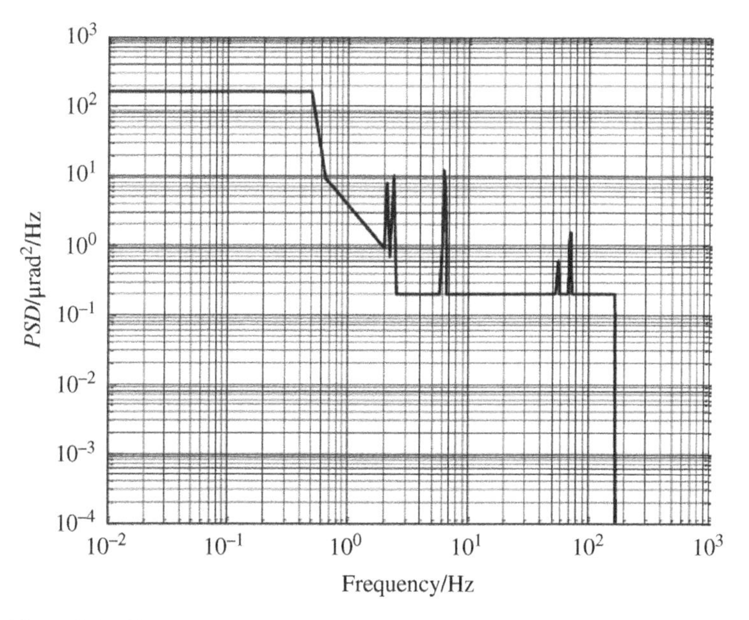

Figure 8.8 The measured power spectrum curve of the LANDSAT-4 satellite.

the information. To the precision aiming control system, so as to achieve precise control of the beam. The CCD light spot detector obtains the light spot image data according to the sampling, and transmits it to the precision sighting subsystem, so as to calculate the incident angle of the light beam, and then the position of the antenna of the opposite terminal can be known, as shown in Figure 8.9.

The noise of the CCD spot detector can be characterized by the noise equivalent angle (NEA), as shown in formulas (8.23a and 8.23b):

$$NEA = \left(\left(S + N_P(Var(R_F) + \Delta t R_T)/S^2\right)N(N+1)/3\right)^{\frac{1}{2}} \tag{8.23b}$$

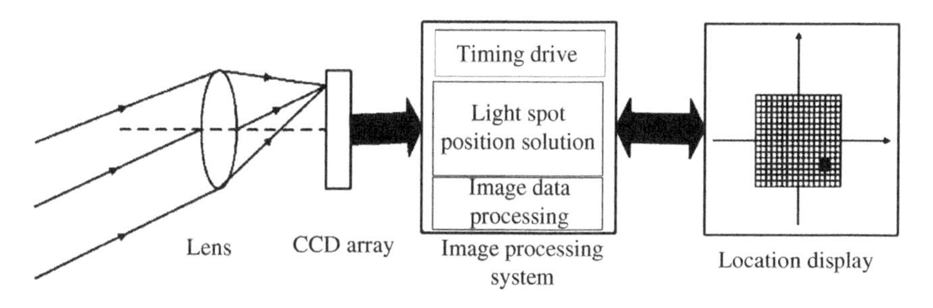

Figure 8.9 Working principle diagram of CCD spot detector.

S is the beam signal received on the detector; Δt is the exposure time; N is the full width at half maximum of the centroid window; N_P is the total number of pixels in the centroid window, equal to $(N+1)^2$; R_F is the fixed noise of a single pixel; $Var(R_F)$ is the equivalent signal corresponding to the fixed noise; R_T is a single pixel background signal, including stray light and dark current. From the formula (8.23b), it can be known that increasing the signal-to-noise ratio can reduce the noise of the CCD detector.

The noise of the CCD detector in the inter-satellite laser communication system mainly comes from three aspects, the first is the inherent noise of the semiconductor optical device itself, and the second is the background light noise. Its specific categories are as follows:

1) Shot noise: In semiconductor devices, in the process of forming a current, the carriers have the characteristic of dispersion, which is the source of shot noise. For a CCD spot detector, the number of signal charges generated in a unit time of beam irradiation is not a stable value, but fluctuates up and down a fixed value, thus forming shot noise. It can be seen from this that shot noise has nothing to do with frequency, so the performance of shot noise on the image can be approximated as Poisson noise.
2) Photon noise: In physics, the definition of the optical power of an optical semiconductor device is not a definite quantity, but the statistical average of the number of photons. Therefore, for the CCD spot detector, even if the input beam power is fixed, the light signal charge collected and stored by the CCD at each moment is random, and this characteristic produces photon noise.
3) Dark current noise: Regardless of other noises, even in the absence of light, the gray value of the image detected by the CCD is not all zero. Because of the thermal motion inside the semiconductor, carriers are generated, and a current is formed at the output, which changes the image data. This is dark current noise. In image data processing, Gaussian noise, Poisson noise, and salt and pepper noise can be used to simulate the above noise.
4) Background light noise: Since the working scene of the inter-satellite laser communication system is in the space environment, with free space and the atmosphere as the link channel, the random scattered photoelectrons generated by stars, planets, and other stars on the CCD detector will cause background radiation, thus forming background light noise; among them, stars are the most numerous and widely distributed stellar light sources in the operating environment of the inter-satellite laser communication system. At any time, random scattered photoelectrons may be generated on the CCD detector, thereby forming a large background spot noise on the CCD detection surface. Therefore, the background light noise mainly refers to the stellar background light. A large number of stellar light sources in space and the reflected light of other stars

caused by the luminescence of stars are likely to enter the window of the CCD detector, which is formed in the CCD spot detector. The spot noise has the characteristics of fixed position, large area, and high brightness. It is generally difficult for the CCD detector of the inter-satellite laser communication system to distinguish it from the effective spot. If it is not processed, it will seriously affect the positioning of the spot center, thus there is a serious deviation in the feedback beam angle information, which seriously degrades the control performance and communication performance of the whole system.

8.3.2.2 Effect of Detector Noise on System Performance

The influence of the noise of the CCD spot detector on the entire inter-satellite laser communication system is that the detected beam image data is disturbed by noise, which affects the processing of the spot image data. In the inter-satellite laser communication system designed in this paper, the CCD is used as the feedback sensor of the precision sighting system. By detecting the laser beam, a light spot is formed and fed back to the precision sighting system, aiming control system closed loop. Therefore, if the spot image obtained by CCD detection is full of noise, it will seriously affect the calculation of the center coordinates of the spot. In order to more clearly demonstrate the influence of the noise of the CCD spot detector on the performance of the inter-satellite laser communication system, the relationship between the spot centroid and the beam incident angle will be briefly introduced below.

As shown in Figure 8.10, it is a schematic diagram of the incident beam imaging on the CCD. When the receiving terminal receives the laser beam, the beam enters the terminal system and reaches the splitter after being reflected by the primary

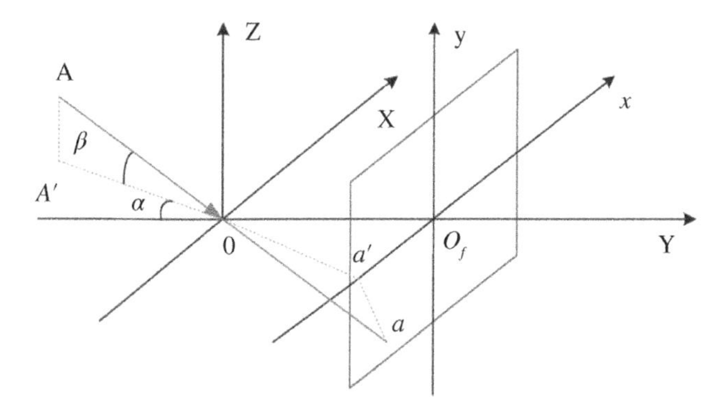

Figure 8.10 Schematic diagram of the relationship between the spot center and the beam incident angle.

mirror and secondary mirror. After a group of optical focusing lenses, the CCD detector is focused and projected into the CCD detector, which is installed on the focal plane of the lens. The CCD detector performs spot detection according to the working principle described above, thereby obtaining a spot image.

Next, the relationship between the incident angle of the beam and the coordinates of the spot center will be analyzed, and a coordinate system OXYZ will be established at the focusing lens, where the OY axis is connected along the line between the center of the lens and the center of the CCD detector window, and the OZ axis is in a plane parallel to the CCD window. The axis is vertical, and the window coordinate system is established in the CCD window plane, where the $O_f x$ axis is parallel to the OX axis, as shown in Figure 8.3, the projection of the incident beam on the OXY plane of the coordinate system OXYZ is $A'O$, the incident beam a is on the detector window plane. The projection point of the coordinate system is s, since the detector is located on the focal plane, there is $OO_f = f$. From the geometric relationship shown in Figure 8.10, we can get:

$$\begin{cases} x_a = \dfrac{f \tan \alpha}{s} \\ y_a = \dfrac{f \tan \beta}{s \cos \alpha} \end{cases} \tag{8.24}$$

Among them, (x_a, y_a) is the coordinate value of the projected light spot of the incident beam a. As can be seen from Figure 8.10, during the communication process of the inter-satellite laser communication system, the APT system makes the incident beam $\overrightarrow{AO}$ as far as possible along the direction $\overrightarrow{OO_f}$ through the control of the coarse sighting control system loop and the fine sighting control system loop, that is, the light spot is detected by the CCD. The coordinates in the plane coordinate system of the monitor window approach (0, 0).

In the control process of the inter-satellite laser communication system, we first obtain the spot coordinates through the CCD detector, and then obtain the incident angle α, β of the beam according to formula (8.24). Because the focal length is of the order of meters, and the pixel pitch s of the CCD detector is generally of the order of microns, which is much smaller than the focal length, the formula (8.24) can be simplified as:

$$\begin{cases} \alpha \approx \tan \alpha \approx \dfrac{s x_a}{f} \\ \beta \approx \tan \beta \approx \dfrac{s \cos \alpha y_a}{f} \approx \dfrac{s y_a}{f} \end{cases} \tag{8.25}$$

During the alignment process, the signal needs to be sent to the coarse sighting control system loop and the fine sighting control system loop of the APT control system. Therefore, the incident angle at the CCD detector should be passed through the primary mirror, secondary mirror, and beam splitter in the terminal optical system. The angle conversion is performed on the installation position of the optical device such as the chip, and the components in the azimuth axis and the pitch axis direction of the terminal system are obtained, so as to facilitate the angle adjustment of the system in these two directions.

8.4 Satellite Platform Vibration Suppression Scheme

In the last chapter, the tracking error of the inter-satellite laser communication system was analyzed, the source of the tracking error was given, and the main influencing factors were pointed out, among which the vibration of the satellite platform was the most influential factor. The vibration of the satellite platform makes the transceiver antenna of the terminal of the inter-satellite laser communication system jitter, which affects the beam emission and incident angle, resulting in the change of the beam distribution of the inter-satellite laser communication system and the increase of the communication BER, and even the communication is interrupted in severe cases. Therefore, it is very necessary to suppress the vibration of the satellite platform [9, 11].

8.4.1 Satellite Platform Vibration Suppression Scheme

At present, the research on the vibration suppression scheme of the satellite platform of the optical communication system at home and abroad mainly focuses on two aspects: passive vibration isolation and active control. We will introduce these two types of suppression schemes separately below.

8.4.1.1 Passive Vibration Isolation

The principle of passive vibration isolation is to put a vibration isolation device between the controlled object and the vibration source, thereby blocking the conduction of vibration and reducing the adverse effects of vibration within a certain range. The key to this solution is the selection of vibration isolation devices. Passive vibration isolation is widely used in occasions where the accuracy requirements are not high and the space is not too limited, because of its simple and reliable structure and low cost.

As shown in Figure 8.11, it is a schematic diagram of two passive vibration isolation structures. The single-stage vibration isolation structure is relatively simple. This passive vibration isolation has a general effect on medium and low frequencies, and is not sensitive to vibration interference and changes in passive objects. Therefore, generally, passive vibration isolation is only used as the primary suppression in the vibration suppression system.

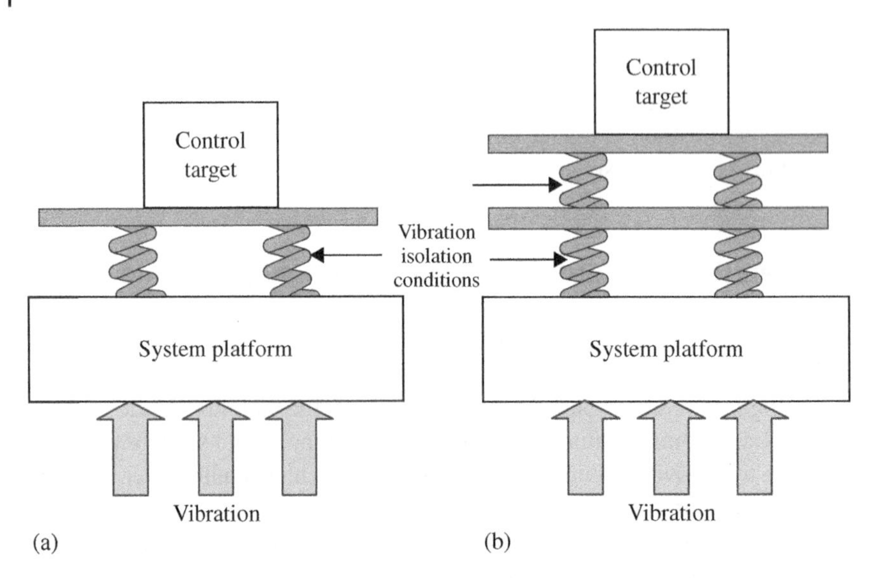

Figure 8.11 Schematic diagram of two passive vibration isolation structures. (a) Single-stage vibration isolation; (b) two-stage vibration isolation.

The two-stage vibration isolation structure is no different from the single-stage structure in principle, but a first-stage vibration isolation system is added on the basis of the single-stage structure. Compared with the single-stage vibration isolation, the two-stage vibration isolation has an improved vibration reduction effect. But it is not very large, and the application of the second-order vibration isolation cannot be when the frequency of the vibration source is on the second-order resonance of the entire system. Moreover, the secondary vibration isolation requires more space, which limits its application in many occasions, especially the application on satellites, and the space limitation has a greater impact [12, 13].

8.4.1.2 Active Control

Although in recent years, for passive vibration isolation, many studies have proposed to change the vibration isolation structure and materials to improve the application of passive vibration isolation, but the temporary effect is not good, and the development prospect is limited, so people began to consider a new solution: active control. Active control is to add an additional actuator to detect vibration information in real time to generate signals with opposite phases and the same amplitude to cancel vibration interference. The structure principle is shown in Figure 8.12.

Inter-satellite laser communication systems at home and abroad generally use the acquisition, aiming, and tracking (APT) structure, and the control structure

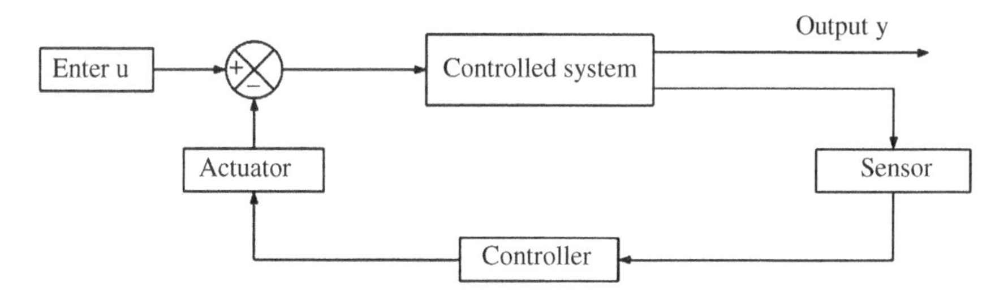

Figure 8.12 Active control structure.

mostly adopts the composite axis structure, which has a certain effect on vibration suppression. In this paper, on the basis of using the APT composite axis control structure, according to the characteristics of the satellite platform vibration, combined with the idea of active control, a suitable suppression algorithm will be designed to achieve the purpose of suppressing the satellite platform vibration.

8.4.2 Feedforward Vibration Suppression Algorithm

8.4.2.1 Influence of Satellite Platform Vibration on Precision Aiming Control System

In order to provide a suitable and effective vibration suppression scheme, and to evaluate the effectiveness of the proposed scheme, before giving the vibration suppression scheme, the impact of the satellite platform vibration on the precision aiming control system should be analyzed in detail, and quantitative data should be given.

In the second chapter, we analyze the composition and structure of the precision aiming system, as well as the structure of the control system. In the previous section, we simulated the vibration of the satellite platform in real time according to the vibration power spectrum of the SILEX platform, which provided the vibration source for our subsequent experimental simulation. Therefore, the control structure diagram of the precision sighting system with the simulated vibration source can be obtained, as shown in Figure 8.13.

Among them, $R(s)$, $V(s)$ are the command and vibration source of the precision sighting control system, $G_c(s)$ is the precision sighting controller, which uses a PID controller, and $G_1(s)$, $G_2(s)$ are the transfer functions of the piezoelectric ceramic (PZT) and fast tilt mirror (FSM) modules, respectively. PZT include a drive amplifier circuit module and a PZT electrical conversion module. The drive amplifier module can be regarded as a proportional amplification link k_a, because PZT can be equivalent to a capacitor, so the PZT electrical conversion module. The transfer function can be expressed as a first-order inertial element $(k_b/T_s + 1)$,

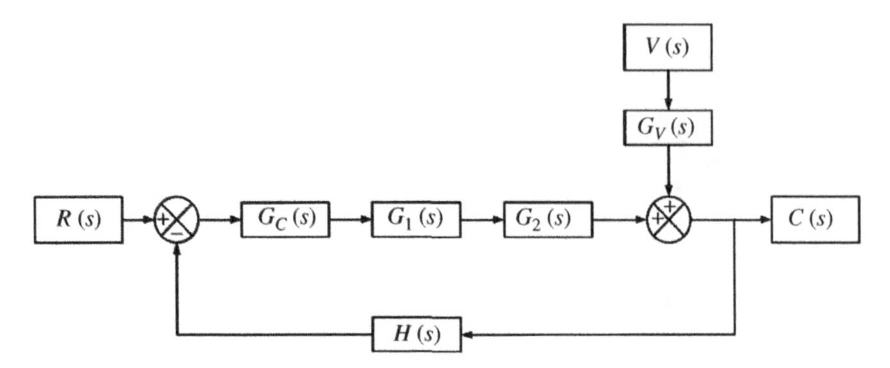

Figure 8.13 Structural diagram of precision aiming control with vibration interference added.

and the transfer function of a fast-tilting mirror can be characterized by a second-order oscillatory element.

So:

$$G_1(s) = \frac{k_a k_b}{T_s + 1} \tag{8.26}$$

$$G_2(s) = \frac{k_c \omega_s^2}{s^2 + 2\xi\omega_s + \omega_s^2} \tag{8.27}$$

where ω_s is the undamped angular frequency of the system, k_c is the proportional coefficient, ξ is the damping ratio, $k_a = 15$, $k_b = 7.76196$, $T_s = 7.2 \times 10^{-5}$, $k_c = 4.732 \times 10^{-7}$, $\xi = 12448.98$, and $\omega_s^2 = 107784759.18$.

Therefore, the transfer function of the control object of the precision sighting system is:

$$\begin{aligned} G(s) &= G_1(s)G_2(s) \\ &= \frac{k_a k_b}{T_s + 1} \cdot \frac{k_c \omega_s^2}{s^2 + 2\xi\omega_s + \omega_s^2} \\ &= \frac{5937.8994}{(0.000072s + 1)(s^2 + 12448.98 + 107784759.18)} \end{aligned} \tag{8.28}$$

Because 0.000072 << 12448.98, the removal of the $T_s + 1$ has little effect on the result. From this, the transfer function of the control object of the precise sighting system can be obtained as:

$$G(s) = \frac{5937.8994}{s^2 + 12448.98 + 107784759.18} \tag{8.29}$$

$G_v(s)$ is the interference channel transfer function, because the vibration source of the satellite platform we use is the simulated angular vibration signal provided in

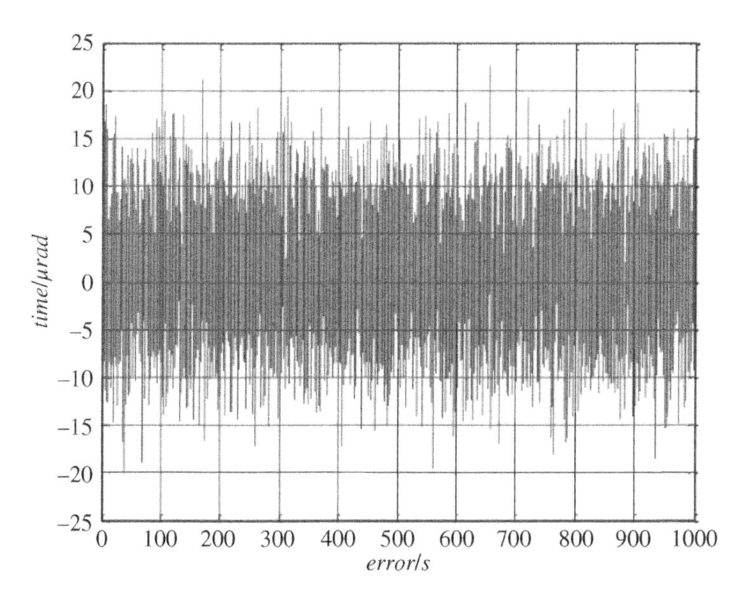

Figure 8.14 Control results of precision sighting system with vibration interference added.

Section 4.3, so $G_v(s)$ can be regarded as 1. $H(s)$ is the transfer function of the spot detector CCD, which can be regarded as a proportional link, and $H(s) = 1$ can be assumed.

According to the structural block diagram of the precision sighting control system shown in Figure 8.13, carry out the matlab simulation, given a step signal with an amplitude of 1, the simulation results as shown in Figure 8.14 can be obtained.

It can be seen from the simulation results that the vibration of the satellite platform brings a large error to the pointing accuracy of the system, and the maximum error exceeds ± 20 μrad. Therefore, it is very necessary to suppress the vibration of the satellite platform.

8.4.2.2 Analysis of Feedforward Vibration Suppression Algorithm

In the application of control system, the more commonly used and effective method for interference suppression is feedforward interference suppression. The standard feedforward control is open-loop control. Unlike PID closed-loop control, it does not require the deviation between the output and input to generate the control variable, but directly suppresses the disturbance through the feedforward loop, so its suppression effect is timelier. But usually in practical applications, feedforward control is basically not used alone, because of the open-loop control characteristics of feedforward control, if it is used alone, it will produce large errors, so it is generally used in combination with closed-loop control [9, 13].

In terms of vibration suppression of optical communication satellite platforms, there are also many studies that apply feedforward control to precision aiming control systems to achieve the effect of suppressing vibration. This paper will analyze the feedforward vibration suppression algorithm in optical communication systems in detail. In order to facilitate the comparison and verification of the vibration suppression algorithm proposed in the next section of this paper, it is to pave the way.

Figure 8.15 shows the block diagram of the precision aiming control system with the feedforward vibration suppression structure added. According to the analysis of the control structure of the precision aiming system in the previous section, $G_1(s)$ and $G_2(s)$ are the transfer functions of PZT and fast tilting mirrors, respectively; $H(s)$ is the transfer function of the spot detector CCD; $V(s)$ is the vibration of the satellite platform; $R(s)$ is the fine tracing control command, here is also a step signal with an amplitude of 1; $G_c(s)$ is the fine sighting controller, using a PID controller; $G_v(s)$ is the interference channel transfer function, because the vibration source of the satellite platform we use is the formula (4.13). The simulated angular vibration signal provided, so $G_v(s)$ can be regarded as 1; $G_{ACC}(s)$ is the transfer function of the accelerometer to detect the vibration of the satellite platform. In the case of a higher sampling frequency, it can be regarded as a proportional link, that is, $G_{ACC}(s) = 1$, $G_{vc}(s)$ is the feedforward vibration suppression controller.

According to Figure 8.15, when the feedforward vibration suppression structure is not used, the total transfer function from the command to the output of the system is:

$$W_R(s) = \frac{G_c(s)G_1(s)G_2(s)}{1 + G_c(s)G_1(s)G_2(s)H(s)} \tag{8.30}$$

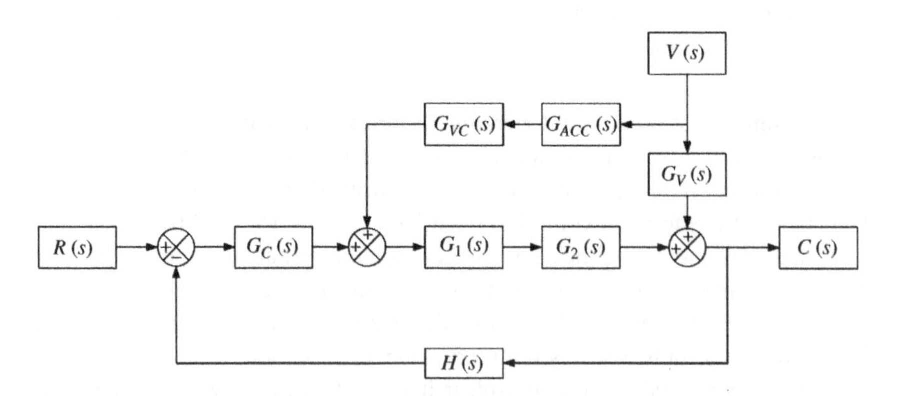

Figure 8.15 Structure block diagram of feedforward vibration suppression.

According to the linear system control theory, if the total transfer function of the system output from vibration interference is 0, the vibration has no effect on the output, so the vibration is completely eliminated, that is,

$$G_{ACC}(s)G_{VC}(s)G_1(s)G_2(s) + G_V(s) = 0 \tag{8.31}$$

By transforming the formula (8.31), the conditions for the precision aiming control system to completely eliminate vibration can be obtained:

$$G_{VC}(s) = \frac{G_v(s)}{G_{ACC}(s)G_1(s)G_2(s)} \tag{8.32}$$

From the analysis in the previous section, it can be seen that the interference channel transfer function $G_v(s)$ represents a rigidly connected rack platform, and the transfer delay is short, and $G_v(s)$ can be simplified as 1. However, the denominator $G_{ACC}(s)G_1(s)G_2(s)$ is a true rational function, and its reciprocal cannot be a true rational function, so the transfer function of the feedforward controller cannot be a true rational function. According to control theory, this is impossible to achieve. In order to realize this feedforward controller, we can transform it. According to the analysis of the control system in Section 4.2, the controller represented by formula (8.32) is three-order, and can also be simplified to two-order, so it can be used in the formula. Adding a third-order denominator in (8.32) is equivalent to adding a filter. After processing, the controller can be physically realized, as shown in formula (8.33). Such a controller can theoretically suppress most vibration disturbances.

$$G_{VC}(s) = \frac{G_v(s)G_s(S)}{G_{ACC}(s)G_1(s)G_2(s)} \tag{8.33}$$

The feedforward controller is designed by formula (8.33), and the simulation analysis of the feedforward vibration suppression algorithm is carried out according to the system parameters obtained from the analysis in the previous section. The simulation results are shown in Figure 8.16.

It can be seen from Figure 8.16 that the feedforward suppression scheme is used, and the error is basically stable below ± 2.5 μrad. However, this is only an ideal situation, and there are many limiting factors in practical application. If the influence of these factors is considered, the error will be much greater than ± 2.5 μrad. Two of the most important limiting factors will be analyzed below [10, 14, 15].

1) Delay of vibration signal measurement. From the analysis in the previous section, it can be known that in the control process, the vibration input of the feedforward channel is measured by the micro-accelerometer and provided in real time, and the vibration interference channel is a rigid structure, and the transfer function can be regarded as "1," then the vibration interference has no effect

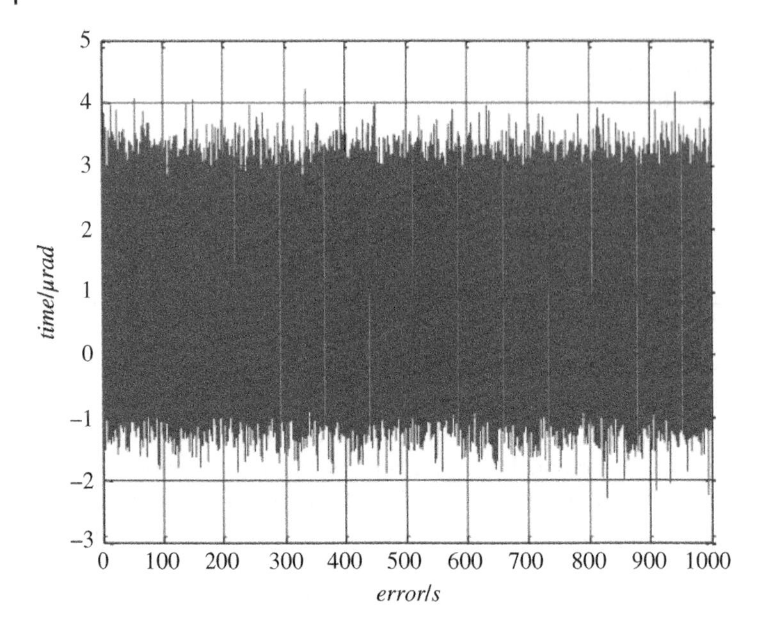

Figure 8.16 Simulation results of feedforward vibration suppression algorithm.

on the time of the two channels of the controlled object. In the same way, the feedforward channel time is obviously longer, and we know that two channels must be canceled to eliminate the influence of vibration, but the current situation is that one channel is delayed by the other channel, so there will be cancellations that are not 0 situation, and even intensify vibration interference. Moreover, it is very difficult to accurately measure the vibration signal, because the signal measured by the accelerometer is the acceleration value, which cannot be directly transmitted to the feedforward controller. Instead, the vibration amount is obtained through two integrations, and then transmitted to the feedforward controller, which in turn increases the runtime of the feedforward channel.

2) Changes in the parameters of the controlled object. In the complex space environment, the parameters of the optical communication terminal system will inevitably be affected and changed. According to the conditions for complete elimination of vibration by the feedforward control provided above, as shown in formula (8.33), if the transfer function parameters of the PZT and the fast mirror change, the control effect of the feedforward vibration suppression algorithm will be greatly affected. It means that the feedforward vibration suppression scheme is very sensitive to changes in system parameters and has poor anti-interference ability.

From the previous analysis, it can be seen that the closed-loop plus feedforward control for vibration suppression can only achieve good results in theory, but in practical application, it faces a large number of limiting factors, including the delay of vibration measurement and the parameters of the controlled object. Changes have a great impact on it, and even increase the interference effect of vibration. So we should reconsider the vibration suppression algorithm.

References

1 Mathur, R.P., Beard, C.I., and Purll, D.J. (1990). Analysis of SILEX tracking sensor performance. *Proceedings of SPIE* 1218: 129–141.

2 Baister, G. and Gatenby, P.V. (1994). Pointing, acquisition and tracking for optical space communications. *Electronics & Communication Engineering Journal* 6: 271–280.

3 Gatenby, P.V. and Grant, M.A. (1991). Optical intersatellite links. *Electronics & Communication Engineering Journal* 3 (6): 280–288.

4 Arnon, S., Rotman, S., and Kopeika, N.S. (1998). Optimum transmitter optics aperture for satellite optical communication. *IEEE Transactions on Aerospace and Electronic Systems* 34 (2): 590–596.

5 Arnon, S. and Kopeika, N.S. (1998). Adaptive bandwidth for satellite optical communication. *IEE Proceedings – Optoelectronics* 145 (2): 109–115.

6 Trent, V., Greene, M., and Hung, S. (1990). Precision pointing error analysis in a satellite optical communication optical system. *IEEE, Proc. Aerospace Conference*, 190–194.

7 Lee, S., Alexander, J.W., and Ortiz, G.G. (2001). Sub-micro radian pointing system design for deep-space optical communications. *Free-Space Laser Communication Technologies XIII* 4272: 104–111.

8 Toyoshima, M. and Araki, K. (2001). In-orbit measurements of short term attitude and vibrational environment on the engineering test staellite VI using laser communication equipment. *Optical Engineering* 40 (5): 827–832.

9 Sudey, J. and Sculman, J.R. (1984). In-orbit measurements of landsat-4 thematic mapper dynamic disturbances. *35th Congress of the International Astronautical Federation*. Lausanne: IAF.

10 Shinhak, L. (2002). Pointing accuracy improvement using model-based noise reduction method. *Free-Space Laser Communication Technologies XIV* 4635: 65–71.

11 Wittig, M., Van Holtz, L., and Tunbridge, D.E.L. (1990). In-orbit measurements of icroaccelerations of ESA's communication satellite LYMPUS. *Free-Space Laser Communication Technologies II* 1218: 205–214.

12 Shan, T.J. and Kailaith, T. (1988). Adaptive algorithm with an automatic gain control feature. *IEEE Transactions on Circuits and Systems* 35: 122–127.

13 Zhang, M., Lan, H., and Ser, W.W. (2001). Cross-updated active noise control system with online secondary path modeling. *IEEE Transactions on Speech and Audio Processing* 9: 598–602.

14 Chan, V.W.S. (2002). Opitcal space communications. *IEEE Journal of Quantum Electronics* 6 (6): 959–975.

15 Shlomi, A., Rotman, S.R., and Kopeika, N.S. (1996). Optimum transmitter optics aperture for free space satellite optical communication as a function of tracking system performance. *Proceedings of SPIE* 2811: 252–263.

9

Inter-Satellite Link Laser Modulation Mode

Broadly speaking, any communication method that uses light as an information transmission medium can be called optical communication. According to the different physical quantities used in the optical domain, the optical modulation format can be divided into four forms of intensity, frequency, phase, and polarization data modulation. At the same time, an optical communication system can be divided into a direct detection system and a coherent system according to the modulation method and the detection method of the receiving end: the former adopts intensity modulation and direct detection; the latter uses the transmitted signal to change after the optical radiation is generated. The amplitude, phase, or frequency of the optical carrier, that is, the external modulation technology, is combined with the local oscillator optical coherent detection scheme [1].

9.1 Block Diagram of Inter-Satellite Link Optical Communication System

The laser communication system applied to the inter-satellite link can adopt either the Intensity Modulation/Direct Detection (IM/DD) system or the coherent optical communication system. From the distance of tens of thousands of kilometers of inter-satellite links, it can be inferred that the optical signal power at the receiving end is weak, and the coherent system will be more sensitive than direct detection. The block diagram of the IM/DD system is shown in Figure 9.1. The transmitting end directly loads the transmitted information onto the optical pulse. The receiving end adopts the direct detection method, which has a simple structure and low cost. The direct detection uses the square-law detection of the photodetector, so only the amplitude information can be detected [2].

Laser Inter-Satellite Links Technology, First Edition. Jianjun Zhang and Jing Li.

Published 2023 by John Wiley & Sons, Inc.

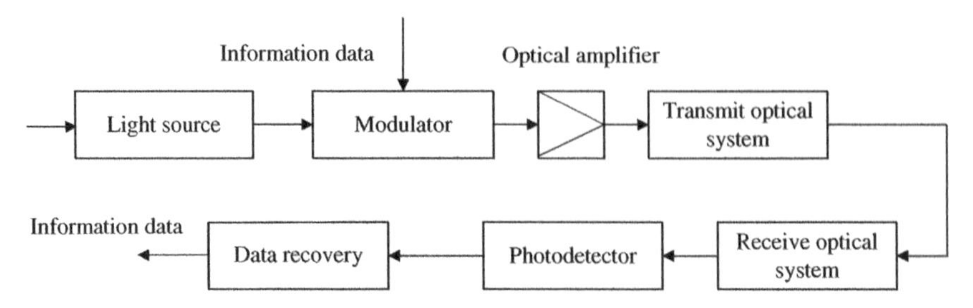

Figure 9.1 IM/DD system composition block diagram.

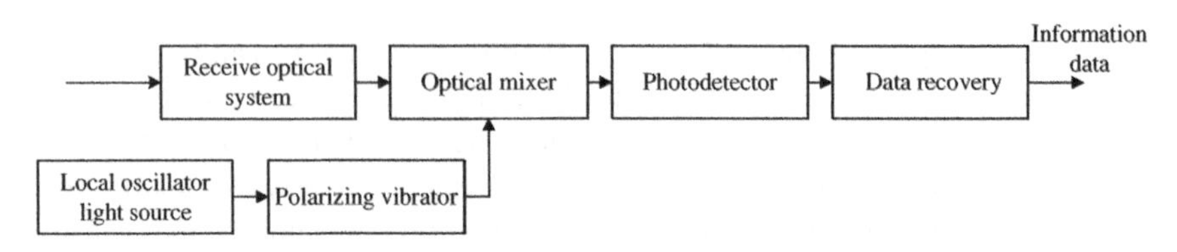

Figure 9.2 Block diagram of the receiving end of the coherent system.

The composition of the transmitting end of the coherent optical communication system is similar to that of the IM/DD system. The difference is that the external modulation method uses the transmitted signal to change the frequency, phase, or amplitude of the optical carrier after the optical radiation is generated. The main difference between the coherent system and the IM/DD system lies in the different detection methods. The block diagram of the receiving end of the coherent system is shown in Figure 9.2. The two are input into the optical mixer, and the photocurrent is output through the action of the photodetector. After passing through the optical mixer, the output IF signal power component contains all the information of the signal light intensity, frequency, or phase, so any modulation method adopted by the transmitter can be retained and reflected in its output. It can be seen that the coherent receiving method has certain versatility in the optical communication system [3].

9.2 Typical Incoherent Optical Modulation (IM/DD)

A variety of modulation methods are suitable for incoherent systems. This section mainly introduces the principles of several IM/DD modulation techniques that are popular in the field of optical communication.

9.2.1 On-Off Key Control

On-off keying (OOK) is the easiest to implement in principle. The input signal of the OOK system is a digital pulse signal. The optical pulse is on or off in each bit time. The "1" bit and the "0" bit are used to encode the time slot with or without the laser pulse, respectively. If the probabilities of the two are equal, the laser needs to work for half of the total duty cycle. The bandwidth required for OOK modulation is inversely proportional to the pulse width, and an OOK signal can be expressed as:

$$s(t) = \sum_{n} a_n g(t - nT_s) \tag{9.1}$$

T_s is the symbol duration, $g(t)$ is the baseband pulse waveform of the duration, and a_n is the level value of the symbol. It can be seen that the maximum bit rate of the system depends entirely on the switching rate of the light source. Since the switching rate of semiconductor lasers is usually high, it is not difficult for an OOK system to provide a certain high-speed bit rate. However, the intensity of the optical pulse depends on the transmission power, and the transmission power decreases with the increase of the modulation rate, so it is difficult to realize long-distance or ultra-high-speed optical communication with OOK modulation.

9.2.2 Pulse Position Modulation

The emission of periodic light pulses at intervals is characteristic of pulse position modulation (PPM), which uses the position of the light pulse to convey information. Each frame of the PPM signal can be divided into $L = 2^M$ time slots with a width of T_s, and the width of the s optical pulse is also T_s, and the position corresponding to the change of the optical pulse represents different code words, and its corresponding information bit is $\log_2 L$. For example, when $M = 2$, the corresponding relationship between OOK modulation and PPM modulation is shown in Figure 9.3.

One of the advantages of the PPM system is that the overhead of the system light source transmission power is significantly reduced. Under the condition of the same optical pulse repetition frequency, it can bring a high data transmission rate with a lower average optical transmission power. At the same time, the anti-interference ability of the PPM system is outstanding, and it is suitable for deep space and submarine laser communication systems [4].

9.2.3 Differential Pulse Position Modulation

Taking single PPM as the prototype, differential pulse position modulation (DPPM) can be obtained after improvement and upgrading, and the number of code groups is not clearly specified. The DPPM signal is to remove the "0" after

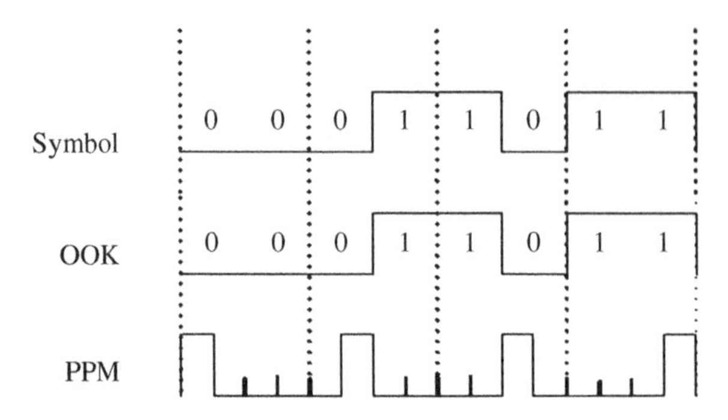

Figure 9.3 PPM modulation timing diagram.

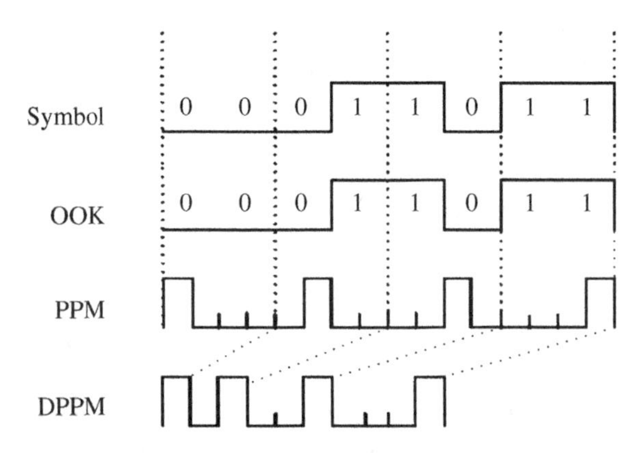

Figure 9.4 DPPM modulation timing diagram.

the "1" time slot in the corresponding PPM signal. Therefore, compared with an L-PPM code group, the information bits transmitted by the two are the same $\log_2 L$. But, when $M = 2$, the schematic diagram of DPPM modulation corresponding to PPM modulation is shown in Figure 9.4.

If a signal transmission rate is given, compared with PPM signal, DPPM signal occupies smaller channel bandwidth and has higher power utilization rate. The requirement for average transmit power is lower than OOK, and the requirement for symbol synchronization is not strict. Because of this, the misjudgment of a time slot not only causes the misjudgment of the current code word, but also the error code will be transmitted to the following code words. It goes without saying that the amount of signal data using DPPM is variable, which limits the range of use of the DPPM system to a certain extent [5].

9.2.4 Digital Pulse Interval Modulation

Digital pulse interval modulation (DPIM) is derived from the pulse interval modulation (PIM) method. PIM is similar to PPM; PPM transmits information through optical pulse time slot position and PIM signal transmits information through the time slot difference between adjacent optical pulses. Each frame of PIM 2^M is still composed of $L = 2^M$ time slots, and the number of time slots between adjacent optical pulses represents every M bits of binary information.

The number of slots contained in each DPIM symbol is variable. A DPIM codeword usually carries a guard slot, and the DPIM signal can be classified according to the presence or absence of the guard slot. The purpose of adding guard time slots is to suppress inter-symbol crosstalk. A DPIM codeword S_k contains $k+2$ time slots, and the optical pulse is located in the initial k time slot, followed by a guard time slot, and finally with empty time slots. Therefore, once the receiving end detects the laser pulse, it can count the following empty time slots and then decrement it by 1. It can be seen that the receiving end of the system can omit symbol synchronization, which reduces the complexity of system implementation. However, if an error occurs in a symbol, even if the symbols before and after it are judged correctly, it may cause false detection [6].

9.2.5 Double Head Pulse Interval Modulation

Double-headed pulse interval modulation (DH-PIM) is an improved PIM modulation method that has been studied more in recent years. According to the name, it can be judged that each DH-PIM symbol can use two different header sequences in the form of H_0 and H_1 and the length of both header sequences are $T_n = [\alpha + 1]$ T_s (α is an integer), but different start pulses increase the modulation method complexity. The number of time slots contained in each symbol information is also different. Each symbol information is composed of a header sequence H_0 and H_1 plus n empty time slots, where the value of n is as follows:

$$n = \begin{cases} k, & k < 2^{M-1} \\ 2^{M-1}k, & k \geq 2^{M-1} \end{cases} \tag{9.2}$$

The number of time slots of the two header sequences is the same, both are $\alpha + 1$, the difference is the number of guard time slots that follow; when $k < 2^{M-1}$, the header sequence takes H_0, and vice versa, when $k \geq 2^{M-1}$, the header sequence takes H_1. The relationship between the two start pulse widths and the number of guard time slots is shown in Table 9.1.

In order to have transition edges within each DH-PIM symbol, take $\alpha = 2$ as an example, take the case that $H_0 = [1\ 0\ 0]$ corresponds to the time slot of the OOK symbol header is "0," and take the case that $H_0 = [1\ 0\ 0]$ corresponding OOK

Table 9.1 Symbol information of different header sequences.

Head pulse	H_0	H_1
Pulse width	$\alpha/2$	α
Number of guard slots	$\alpha/2+1$	1

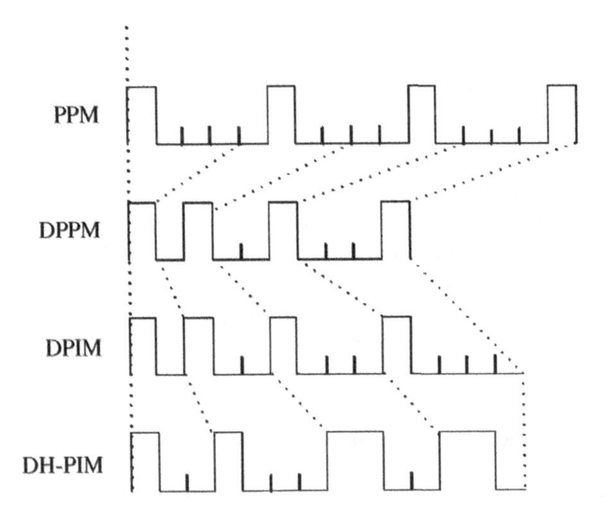

Figure 9.5 The corresponding timing diagram of several modulation methods under the same string of symbols.

symbol header time slot is “1”. Figure 9.5 is the timing diagram of DPIM and DH-PIM modulation corresponding to Figure 9.4.

9.3 Coherent Optical Communication Modulator and Modulation Principle

The basic principle of coherent light modulation: a high-performance laser carrier generated by a light source is modulated by a light modulator controlled by an external electric field, and then the modulated light wave is transmitted. The signal light at the receiving end is mixed with the local oscillator light, down-converted to the intermediate frequency, and then demodulated to obtain the baseband output [7, 8].

9.3.1 Optical Modulator

The coherent optical modulator modulates the electrical signal carrying the data onto the optical carrier generated by the laser to realize electrical–optical conversion. Commonly used light modulators are generally based on the electro-optic effect of the material – the refractive index of the electro-optic material changes with the applied external voltage to achieve optical frequency, phase, and amplitude modulation.

The working mechanism of the phase modulator is precisely based on the electro-optic effect to change the propagation speed and phase of the light wave. The electro-optic effect found that the refractive index of the crystal material has a complex relationship with the external electric field E, which can be approximated as $\Delta n \sim (R|E| + \gamma|E|^2)$. It can be seen that $R|E|$ and E have a linear relationship, and $\gamma|E|^2$ and E have a square relationship. The former linear relationship is called the Pockels effect, and the latter square relationship is called the Kerr effect. The Pockels effect is the principle on which the optoelectronic phase modulator came out. The structure of the optical phase modulator is shown in Figure 9.6.

When only the first-order Pockels effect is considered, the change in phase can be viewed as a linear function of the input drive voltage $u(t)$:

$$\varphi_{PM}(t) = \frac{2\pi}{\lambda} \cdot k\Delta n_{eff} L \cdot u(t) \tag{9.3}$$

$\varphi_{PM}(t)$ is the modulation coefficient of the phase modulator, L is the interaction length between the light and the electrode, λ represents the wavelength of the input light, and n_{eff} represents the change of the effective refractive index. At the same time, the half-wave voltage V_π defines the main characteristic parameters

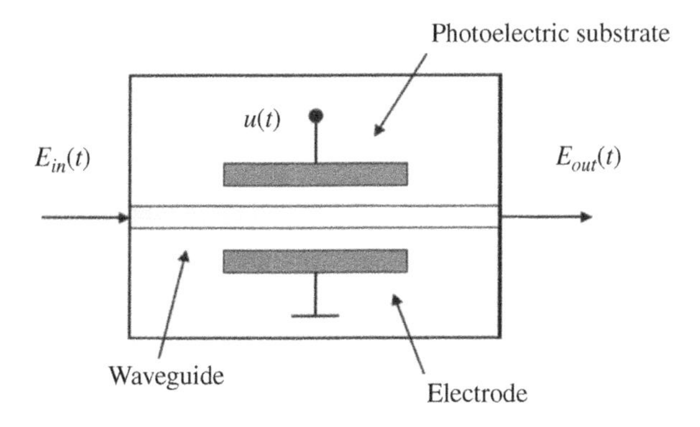

Figure 9.6 $LiNbO_3$ crystal integrated optical phase modulator structure.

of a modulator, that is, the voltage required when the modulator phase is π reversed:

$$V_\pi = \frac{\lambda}{2k\Delta n_{eff}L} \tag{9.4}$$

From this, the relationship between the modulation coefficient $\varphi_{PM}(t)$ and $u(t)$ and V_π is:

$$\varphi_{PM}(t) = \frac{u(t)}{V_\pi}\pi \tag{9.5}$$

9.3.2 Coherent Optical Communication Modulation Format

Due to the large data transmission capacity and long distance of the inter-satellite link, high bit rate data transmission capability and high spectral efficiency are required. The multidimensional and multi-order modulation format can reduce the baud rate of data transmission by transmitting multi-bit data information in each symbol, thereby achieving high spectral efficiency transmission. Therefore, multi-order modulation formats can be considered for inter-satellite links.

9.3.2.1 Binary Phase Shift Keying

BPSK is a binary phase-shift modulation format. From the research status at home and abroad in the previous chapter, it can be seen that it is the most widely used coherent modulation format for inter-satellite laser links. The Mach–Zehnder modulator (MZM) is an optical modulator based on the interference principle, which is constructed by combining two phase modulators in parallel on a $LiNbO_3$ crystal substrate. A pair of parallel strip waveguides are fabricated on the $LiNbO_3$ substrate, and a 3 dB Y-shaped branch waveguide is connected at both ends. The structure of the MZM modulator is shown in Figure 9.7.

The input beam E_m passes through a 3 dB Y-type branch waveguide and is coupled into two parallel waveguides with the same structural parameters under the driving of the upper and lower arms. Convert to intensity modulation. The MZM transfer function can be expressed as:

$$E_{out} = E_{in} \cdot \frac{1}{2}\left(e^{j\varphi_1(t)} + e^{j\varphi_2(t)}\right) \tag{9.6}$$

$\varphi_1(t)$ and $\varphi_2(t)$ denote the phase shifts of the upper and lower arms of the MZM, respectively. According to the principle of the phase modulator, the phase shift of the upper and lower arms can be expressed as:

$$\varphi_1(t) = \frac{u_1(t)}{V_{\pi 1}}\pi,\ \varphi_2(t) = \frac{u_2(t)}{V_{\pi 2}}\pi \tag{9.7}$$

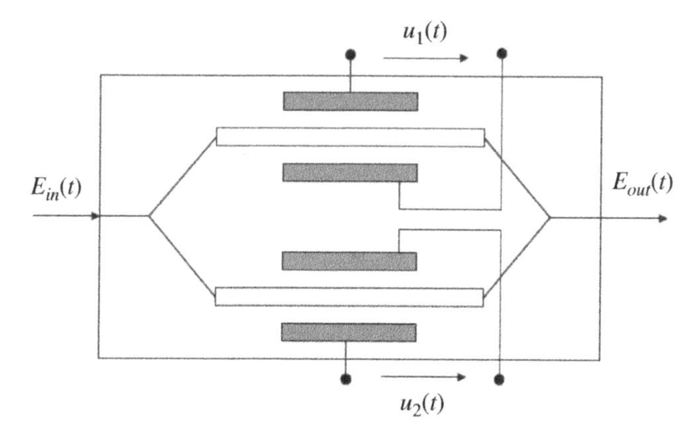

Figure 9.7 MZM modulator structure.

$V_{\pi 1}$ and $V_{\pi 2}$ represent the half-wave voltage of the upper and lower arms, respectively, $u_1(t)$ and $u_2(t)$ represent the external voltages of the upper and lower arms, respectively, including the RF driving voltage and the DC bias voltage. The modulated output E_{out} signal can be expressed as:

$$\begin{aligned} E_{out} &= E_{in}\frac{1}{2}[\cos\varphi_1(t) + \cos\varphi_2(t) + j[\sin\varphi_1(t) + \sin\varphi_2(t)]] \\ &= E_{in}\frac{1}{2}\left[\cos\left(\frac{\varphi_1(t)+\varphi_2(t)}{2}\right)\cos\left(\frac{\varphi_1(t)-\varphi_2(t)}{2}\right)\right. \\ &\quad \left. + j\sin\left(\frac{(\varphi_1(t)+\varphi_2(t))}{2}\right)\cos\left(\frac{(\varphi_1(t)-\varphi_2(t))}{2}\right)\right] \\ &= E_{in}\cos\left(\frac{\varphi_1(t)-\varphi_2(t)}{2}\right)\exp\left(j\frac{\varphi_1(t)+\varphi_2(t)}{2}\right) \end{aligned} \tag{9.8}$$

It can be seen that $\varphi_1(t) = \varphi_2(t)$, the phase shift of the upper and lower arms was the same, and the MZM was working in the push–push mode, and the signal was phase-modulated; when $\varphi_1(t) = -\varphi_2(t)$, the phase shifts of the upper and lower arms were opposite, and the MZM was working in the push–pull mode. Intensity modulation is performed. Assuming that the MZM is biased at 0 and the amplitude of the driving signal fluctuation is $2V_\pi$, a BPSK signal $E_{out} = E_{in}e^{j\pi b_k}$ can be generated, where b_k is the input data bit sequence.

9.3.2.2 Quaternary Phase Shift Keying

QPSK is one of the most sought-after multi-ary modulation formats in the field of high-speed optical communication. The QPSK signal has a symbol rate of one-half the total bit rate and transmits four phase shifts (invariant amplitude) in a

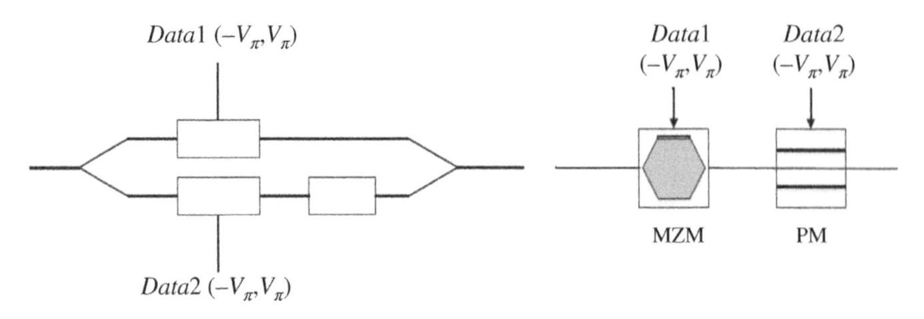

Figure 9.8 Parallel and cascaded QPSK modulation structures.

quaternary phase shift modulation format. The required QPSK signal can be generated by different combinations of the two MZMs. The structure of QPSK modulation is shown in Figure 9.8, which can be divided into parallel and series. In the parallel scheme of the double-balanced structure, both MZMs are biased at 0, and the amplitude of the driving signal fluctuation is $2V_\pi$, so that two BPSK signals are generated. The two BPSK signals are:

$$E_1 = E_{in}e^{j\pi c_k},\ E_2 = E_{in}e^{j\pi(d_k + 0.5)} \tag{9.9}$$

Among them, c_k and d_k are the two channels of input modulated data, and then the two channels of orthogonal BPSK signals are coupled to generate QPSK signals, so the end output is:

$$E_{out} = E_{in}\left(e^{j\pi c_k} + e^{j\pi(d_k + 0.5)}\right) \tag{9.10}$$

In the modulation block diagram of the tandem structure, the input signal passes through a MZM, and then forms a QPSK signal through the action of a phase of $\pi/2$ modulator (PM). At this time, the output signal expression is $E_{out} = E_{in}e^{j\pi(c_k + 0.5d_k)}$, c_k and d_k are the two-channel data modulated by the input.

It can be seen from the Figure 9.8 that it is not difficult to form a QPSK signal in these two ways, and it can be easily realized under the existing hardware conditions. However, in practical applications, the parallel structure is mostly used, because the PM in the series structure will be driven by the current. The jitter of the optical phase is unstable, which will affect the performance of the QPSK modulator in some real-world applications.

9.3.2.3 8PSK

The 8PSK transmitter consists of one MZM and two PMs, as shown in Figure 9.9. c_k, d_k, and e_k are three-way data modulated by the input. By observing the QPSK modulator in the series structure, the series 8PSK signal is equivalent to adding a

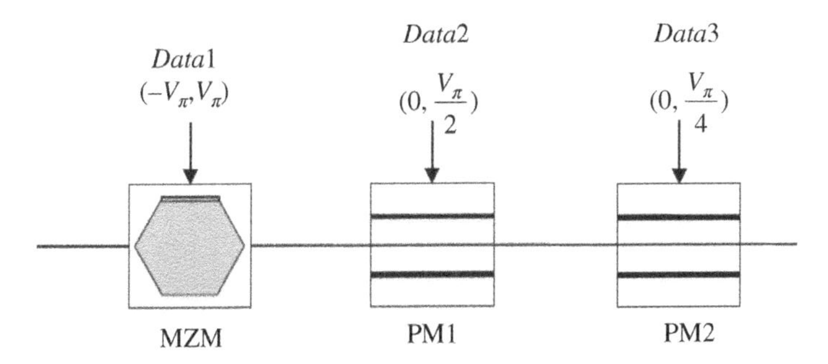

Figure 9.9 Cascaded 8PSK modulation structure.

level of $(0, \pi/4)$ of PM. MZM, PM1, and PM2 provide phase modulation $0/\pi$, $0/0.5\pi$, and $0/0.25\pi$, respectively, and the outputs after passing through the three modulators are:

$$E_1 = E_{in}e^{j\pi c_k},\ E_2 = E_{in}e^{j\pi(c_k + 0.5d_k)},\ E_3 = E_{in}e^{j\pi(c_k + 0.5d_k + 0.25e_k)} \tag{9.11}$$

The effect of adding one stage of PM to generate 8PSK after the tandem QPSK is not particularly ideal. In this method, the phase shift change will cause undesired signal chirp. The structure shown in Figure 9.10 is a new idea, which generates 8PSK signals from four parallel MZMs (QPMZMs). In this structure, each MZM bias point is set to 0, each arm generates a BPSK signal, and the optical phase bias voltages of these arms are different (respectively, $n\pi/4$, $n = -1, 0, 1, 2$). When the input-modulated three-way data is encoded, the upper and lower QPSK signals are successfully obtained by relying on $[MZM - I_1,\ MZM - Q_1]$ and $[MZM - I_2,\ MZM - Q_2]$ to form different combinations. The optical phases of the upper and

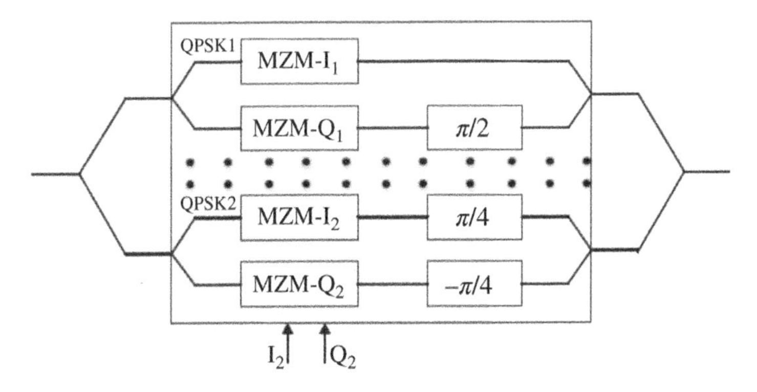

Figure 9.10 8PSK modulation of QPMZM structure.

lower two-way signals are biased at $\pi/4$, and the required 8PSK signal is coupled. The signal formed by this four parallel MZM structures can avoid the influence of frequency chirp. Through the principle of series and parallel 8PSK structure diagrams, it is not difficult for us to think that there should be a series–parallel combination structure diagram.

9.3.2.4 8QAM

Quadrature Amplitude Keying (QAM) is a joint keying where the phase and amplitude of the signal are modulated as different independent parameters. Figure 9.11a is a serial structure 8QAM modulation scheme under all-optical conditions, consisting of a double-balanced and a *PM*. $\pi/4$ Biased double-balanced MZMs are similar to QPSK modulators, in that the bias points are all 0; however, the amplitude of the drive signal fluctuation of is only $0.7V_\pi$. The constellation diagram of the superposition output of the two signals is shown in Figure 9.11b, and

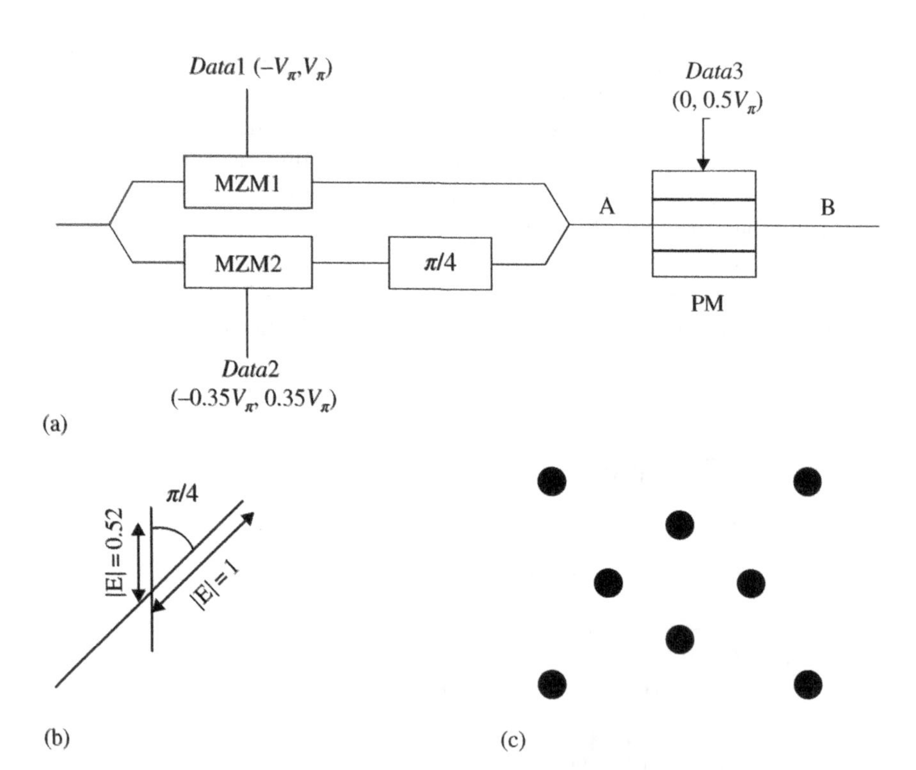

Figure 9.11 8QAM modulation structure. (a) All-optical 8QAM modulation structure diagram; (b) 4QAM schematic diagram; (c) 8QAM constellation diagram.

then rotated by the phase modulator of $\pi/2$, the 8QAM constellation diagram shown in Figure 9.11c can be obtained.

9.4 Comparison of Communication Performance of Laser Modulation Schemes

This section mainly compares the communication performance of several typical laser modulation schemes from three aspects: average transmit power, transmission bandwidth requirement, and bit error rate [9].

9.4.1 Average Transmit Power

Assuming that the same codeword is sent in the following modulation modes, calculate the required average transmit power under the condition of the same bit rate.

9.4.1.1 OOK

For non-return-to-zero (NRZ) OOK, this paper assumes that when the symbol "1" is transmitted, the luminous power of the light source is P_1, then when the symbol "0" and the symbol "1" in the transmitted valid information have the same probability of occurrence, the average luminous power is $P_1/2$.

9.4.1.2 PPM

Assuming that the average transmitted optical power of OOK (NRZ) modulation is P_{OOK}, in the PPM signal, the high-level pulse only occupies one clock cycle in each PPM symbol. For an M-bit codeword, there are a total of 2^M time cycles, but there is still only one high-level optical pulse period that can emit light. Therefore, the average transmit power of the PPM signal is:

$$P_{PPM} = \frac{P_1}{2^M} = \frac{P_{OOK}}{2^{M-1}} \tag{9.12}$$

9.4.1.3 DPPM

For an M-bit codeword, the symbol of each DPPM signal consists of $(2^M + 1)/2$ clock cycles, but the high-level pulse of the light only occupies one clock cycle, so the average transmit power of the DPPM signal is:

$$P_{DPPM} = \frac{2P_1}{(2^M + 1)} = \frac{4P_{OOK}}{(2^M + 1)} \tag{9.13}$$

9.4.1.4 DPIM

Similar to DPPM, the symbol of each DPIM signal consists of $(2^M + 3)/2$ clock cycles, so the average transmit power of DPIM signal is:

$$P_{DPIM} = \frac{2P_1}{(2^M + 3)} = \frac{4P_{OOK}}{(2^M + 3)} \tag{9.14}$$

9.4.1.5 DH-PIM

The special feature of DH-PIM signal is that if the header sequence H_0 and H_1 and the probability of occurrence are the same, the header sequence occupies 1.5 cycles in all pulse cycles of each DH-PIM symbol, and each pulse occupies $\alpha/2$ time slot; thus, the average time slot length $\overline{L} = 3\alpha/4$ of the head sequence pulse can be calculated, so the average transmit power is:

$$P_{DH-PIM} = \frac{3\alpha P_1}{2(2^{M-1} + 2\alpha + 1)} = \frac{3\alpha P_{OOK}}{2^{M-1} + 2\alpha + 1} \tag{9.15}$$

9.4.1.6 Coherent PSK

In OOK modulation, the luminous power when the light source sends the original information "1" is P_1, then the average luminous power is $P_1/2$, and because the coherent PSK modulation is a constant envelope modulation, the average luminous power of the coherent PSK signal is:

$$P_{BPSK} = \frac{2^M}{2^M} P_1 = 2P_{OOK} \tag{9.16}$$

9.4.2 Transmission Bandwidth

For a communication system, it can generally be considered that the main lobe width of the power spectral density is equivalent to its bandwidth, so its bandwidth is a $\sin c$ function. Since the time slot width of the transmitted wireless optical pulse is extremely narrow, the bandwidth of the signal can be approximately estimated by the inverse of its time slot width. Assuming that the information transmission bit rate is constant and the modulation order is M, the bandwidth requirements of several modulation methods are compared below [10].

9.4.2.1 PPM

Assuming that the information bit rate of the OOK signal is b_R and the average symbol length is T_{OOK}, then the bandwidth occupied by the OOK modulation is $B_{OOK} = R_b$, then the bandwidth required by the PPM signal is:

$$B_{PPM} = \frac{2^M}{M} R_b \tag{9.17}$$

9.4.2.2 DPPM

Compared with the OOK signal, the average symbol length of the DPPM signal is $[2M/(2^M + 1)]T_{OOK}$, so its transmission bandwidth can be expressed as:

$$B_{DPPM} = \frac{2^M + 1}{2M} R_b \tag{9.18}$$

9.4.2.3 DPIM

Compared with the OOK signal, the average symbol length of the DPIM signal is $[2M/(2^M + 3)]T_{OOK}$, so its transmission bandwidth can be expressed as:

$$B_{DPIM} = \frac{2^M + 3}{2M} R_b \tag{9.19}$$

9.4.2.4 DH-PIM

Similarly, the average symbol length of DH-PIM signal is $[2M/(2^{M-1} + 2\alpha + 1)]$ T_{OOK}, which is an integer, corresponding to different header sequences, so its transmission bandwidth can be expressed as:

$$B_{DH-PIM} = \frac{2^{M-1} + 2\alpha + 1}{2M} R_b \tag{9.20}$$

9.4.2.5 Coherent PSK

When the bit rate of information transmission is the same, the spectral characteristics of BPSK signal are similar to OOK. MPSK signal symbol can be regarded as the sum of two specific MASK signal symbols, so its bandwidth is the same as that of MASK signal, so coherent PSK modulation signal transmission bandwidth can be expressed as:

$$B_{BPSK} = B_{OOK} = R_b \tag{9.21}$$

9.4.3 Bit Error Rate

The bit error rate (bit error ratio, BER) reflects the reliability of a communication system, that is, the accuracy of data transmission within a valid period of a communication system. During the entire transmission process of any signal from the transmitter to the receiver, it will inevitably be subjected to various interferences in the transmission channel environment, which may cause bit errors at the receiver.

For the convenience of comparison, the standard is unified in all calculations below. It is assumed that white Gaussian noise with mean value of 0 and variance of σ_n^2, the peak power of the signal entering the receiver decider is S_t, and the bandwidth of the receiver is relatively large. In this case, the restored values of the valid information "1" and "0" are, respectively, $\sqrt{S_t + n(t)}$ and $\sqrt{n(t)}$.

9.4.3.1 OOK

Before calculating the bit error rate, the effective decision threshold must be assumed in advance. The occurrence of bit errors can be divided into two cases: misjudging the symbol "1" as the symbol "0" and misjudging the symbol "0" as the symbol "1." Using P_{10} and P_{01} to represent the probabilities when these two situations occur, there are:

$$P_{01} = \frac{1}{2}\left\{1 - erf\left[\frac{b}{\sqrt{2\sigma_n^2}}\right]\right\},\ P_{10} = \frac{1}{2}\left\{1 + erf\left[\frac{b - \sqrt{S_t}}{\sqrt{2\sigma_n^2}}\right]\right\} \tag{9.22}$$

where $erf(x) = (2/\sqrt{\pi})\int_0^x \exp(u^2)du = 1 - erf(x)$. It is assumed that at the transmitting end, the probability of transmitting symbols "1" and "0" is equal, which is 1/2. From this, the effective decision threshold $b = \sqrt{S_t}/2$ can be obtained, and finally the bit error rate of OOK modulation can be expressed as:

$$P_{OOK} = \frac{1}{2}P_{01} = \frac{1}{2}P_{10} = \frac{1}{2}erfc\left(\frac{\sqrt{S_t/2\sigma_n^2}}{2}\right) \tag{9.23}$$

9.4.3.2 PPM

The bit error rate of PPM modulation can be expressed as:

$$P_{PPM} = \frac{1 + erf\left[(b - \sqrt{S_t})/\sqrt{2\sigma_n^2}\right] + (2^M - 1)\left[1 - erf\left(b/\sqrt{\sigma_n^2}\right)\right]}{2^{M+1}} \tag{9.24}$$

Taking the derivative of the above formula, and then setting the derivative to 0, we can obtain $b = \left[2\sigma_n^2 \ln(l) + S_t\right]/2\sqrt{S_t}$, $l = 2^M - 1$.

9.4.3.3 DPPM

The bit error rate of DPPM modulation can be expressed as:

$$P_{DPPM} = \frac{1 + erf\left[(b - \sqrt{S_t})/\sqrt{2\sigma_n^2}\right] + (2^M - 1)/2\left[1 - erf\left(b/\sqrt{\sigma_n^2}\right)\right]}{2^{M+1}} \tag{9.25}$$

Taking the derivative of the above formula, and then setting the derivative to 0, we can obtain $b = \left[2\sigma_n^2 \ln(l) + S_t\right]/2\sqrt{S_t}$, $l = (2^M - 1)/2$.

9.4.3.4 DPIM

Similarly, the bit error rate of DPIM modulation can be expressed as:

$$P_{DPIM} = \frac{1 + erf\left[\left(b - \sqrt{S_t}\right)/\sqrt{2\sigma_n^2}\right] + (2^M + 1)/2\left[1 - erf\left(b/\sqrt{\sigma_n^2}\right)\right]}{2^{M+3}} \tag{9.26}$$

Taking the derivative of the above formula, and then setting the derivative to 0, we can obtain $b = \left[2\sigma_n^2 \ln(l) + S_t\right]/2\sqrt{S_t}$, $l = (2^M + 1)/2$.

9.4.3.5 DH-PIM

The bit error rate of DH-PIM modulation can be expressed as:

$$P_{DH-PIM} = \frac{(3\alpha/2)\left(1 + erf\left[\left(b - \sqrt{S_t}\right)/\sqrt{2\sigma_n^2}\right]\right) + (4L_m - 3\alpha)/2\left[1 - erf\left(b/\sqrt{\sigma_n^2}\right)\right]}{4L_M} \tag{9.27}$$

Taking the derivative of the above formula, and then setting the derivative to 0, we can obtain $b = \left[2\sigma_n^2 \ln(l) + S_t\right]/2\sqrt{S_t}$, $L_m = (2^M + \alpha + 2)/3\alpha$.

9.4.3.6 BPSK

The bit error rate of BPSK modulation can be expressed as:

$$P_{BPSK} = \frac{1}{2} erfc\left(\sqrt{\frac{S_t}{2\sigma_n^2}}\right) \tag{9.28}$$

9.4.4 Summary

Most of the satellite laser link communication systems that are in the experimental stage or have been formed use incoherent modulation technology or the simplest BPSK modulation method in the coherent system. Due to the incoherent transmission power and amplitude detection methods, the sensitivity of the receiver is not enough, which limits the improvement of the communication rate. At the same time, the data communication rate of the existing coherent system link is much higher than that of the incoherent system. Most of the existing noncoherent systems use OOK and PPM modulation. From the simulation results, it can be found that the former is simple to implement and occupies a small bandwidth. The disadvantage is that the bit error rate performance is not ideal and the power utilization rate is low. The latter optimizes power utilization and bit error rate performance, but bandwidth utilization is not ideal. In the actual inter-satellite application scenario, the thermal noise of the IM/DD system is limited, coupled with background light interference, making it difficult to approach the theoretical limit of detection sensitivity, and the actual performance is at least 20 dB worse than the theoretical limit. The early inter-satellite link communication system can meet the requirements of short-to-medium-distance, medium-low rate communication in a noncoherent way. For long-distance, high-dynamic navigation satellites and future high-precision, high-speed satellite communication requirements, coherent system is undoubtedly a better choice [11, 12].

The advantages of the inter-satellite laser link coherent communication system can be summarized as follows: under the same environment, the detection sensitivity of the receiving end is significantly improved; with the improvement of the receiving sensitivity, the hardware level will also be improved, such as the diameter of the transceiver antenna and the size of the communication machine and quality reduction; by increasing the local oscillator optical power generated by the local oscillator light source, the receiving end can reach the shot noise limit, thereby appropriately reducing the requirements for the transmitting power; flexible and diverse modulation methods, such as BPSK, QPSK, or multi-order formats such as 16QAM; it has strong wavelength selectivity, that is, narrow-band filtering characteristics: the extremely narrow coherent receiver bandwidth is compared with the optical frequency, and the interference of broadband background light can be effectively suppressed by selecting the appropriate optical frequency tracking technology.

References

1 Wuchenich, D.M.R., Mahrdt, C., Sheard, B.S. et al. (2014). Laser link acquisition demonstrationfor the GRACE follow-onmission. *Optics Express* 22 (9): 11351–11366.

2 Rodriguea, I. and Garcia, C. (2011). Inter-satellite links for satellite autonomous integrity monitoring. *Advancesin Space Research* 47: 197–212.

3 Han, S., Gui, Q., and Li, J. (2011). Analysis of the connectivity and robustness of inter-satellite links in aconstellation. *Science China - Physics Mechanics & Astronomy* 54 (6): 991–995.

4 Nielsen, T.T. and Oppenhaeuser, G. (2002). In orbit test result of an operational optical inter-satellite link between ARTEMIS and SPOT4, SILEX. *Free-Space Laser Communication Technologies XIV* 4635: 1–15.

5 Nakamura, N., Kamio, Y., Lu, G.-W et al. (2007). Ultimatelinewidth-tolerant 20Gbps QPSK-homodyne transmission using a spectrum-sliced ASE lightsource. *Optical Fiber Communications Conference, Opt Soc, America*, 25–30.

6 Huang, S. and Wang, L. (2003). Exact evaluation of laser linewidth requirements for optical PSKhomodyne communication systems with balanced OPLL receivers. *Lightwave Technology* 14 (5): 661–664.

7 Vilnrotter, V., Lau, C.-W., and Srinivasan, M. (2004). Optical array receiver for deep-space communication. *Free Space Laser Communication Technologies XVI* 34 (5): 981–988.

8 Wen, K., Zhao, Y., Gaoet, J. et al. (2015). Design of a coherent receiver based on InAs electron avalanche photodiode for free-space optical communications. *IEEE Transactions on Electron Devices* 6 (62): 1932–1938.

9 Atef, M., Swoboda, R., and Zimmermann, H. (2012). 1.25 Gbit/s over 50 m step-index plastic optical fiber using a fully integrated optical receiver with an integrated equalizer. *Journal of Lightwave Technology* 1 (30): 118–122.

10 Robert, L. and Berry, S. (2007). Homodyne BPSK-based optical inter-satellite communicationlinks. *Proceedings of the SPIE* 6457: 31–37.

11 Abusali, P.A.M., Tapley, B.D., and Schutz, B.E. (2003). Autonomous navigation of global position system satellites using cross-link. *Journal of Guidance, Control, and Dynamics* 21 (2): 321–332.

12 Li, H., Li, J., and Jiao, W. (2010). Analyzing perturbation motion and studying configuration maintenance strategy for compass-M navigation constellation. *Journal of Astronautics* 31 (7): 1756–1761.

Index

Laser Inter-Satellite Links Technology, First Edition. Jianjun Zhang and Jing Li.

Published 2023 by John Wiley & Sons, Inc.

Printed and bound by CPI Group (UK) Ltd, Croydon, CR0 4YY

16/04/2025

14658582-0001